Drugs and the Eye

Drugs and the Eye

Second Edition

Approved by the British College of Ophthalmic Opticians (Optometrists)

Janet Vale, MSc, MPS
Lecturer in Ocular Pharmacology
University of Manchester

B. Cox, MSc, PhD, DSc, MPS
Honorary Reader in Pharmacology
University of Manchester

Butterworths
London Boston Durban Singapore Sydney Toronto Wellington

First published, 1978
Reprinted 1978
Second edition 1985

British Library Cataloguing in Publication Data

Vale, Janet
Drugs and the eye.—2nd ed.
1. Eye—Diseases and defects—Chemotherapy
2. Ocular pharmacology
I. Title II. Cox, B. (Barry) III. British College of Ophthalmic Opticians (Optometrists)
617.7′061 RE994

ISBN 0-407-00319-3

Library of Congress Cataloguing in Publication Data

Vale, Janet
Drugs and the eye.—2nd ed.

Bibliography: p.
Includes index.
1. Ocular pharmacology I. Cox, B. (Barry), 1937– . II. British College of Ophthalmic Opticians (Optometrists) III. Title. [DNLM: 1. Eye–drug effects. 2. Eye Diseases–drug therapy WW 166 V149d]
RE994.V34 1985 617.7′061 84-23238

ISBN 0-407-00319-3

Photoset by Butterworths Litho Preparation Department
Printed and bound by Thomson Litho Ltd., East Kilbride, Scotland

Preface to the second edition

Since the publication of the first edition there have been changes and developments which have necessitated the updating of certain chapters. The most notable is the introduction of the Medicines Act replacing the Pharmacy and Poisons Act. The chapter dealing with the legal aspects of sale and supply of drugs has thus been completely rewritten to take account of this. Other areas which have been substantially rewritten deal with the care of contact lenses, reflecting the changed approach to soft contact lenses, and the section on the problem of drug-induced ocular side-effects. Here an attempt has been made, not to list every drug reported to cause such effects, but to discuss the relevance and scope of the problem. The format of the other chapters remains basically as the first edition, with additional information being added as required to bring the topic up to date. The order in which drugs are discussed has been changed in some chapters to reflect more closely the current pattern of usage.

It is hoped that these changes will ensure that the book still meets the aims set out in the first edition, namely of being a working guide for qualified ophthalmic opticians, a source of information for students of ophthalmic optics and one of interest for students of related topics.

M.J.V.

Contents

1

Anatomy and physiology

A horizontal section through the eye is shown in *Figure 1.1*. There are three layers which enclose the transparent media through which light passes before reaching the retina.

The outer layer is protective in function. It is predominantly white in colour and opaque (the sclera) with a transparent anterior portion (the cornea).

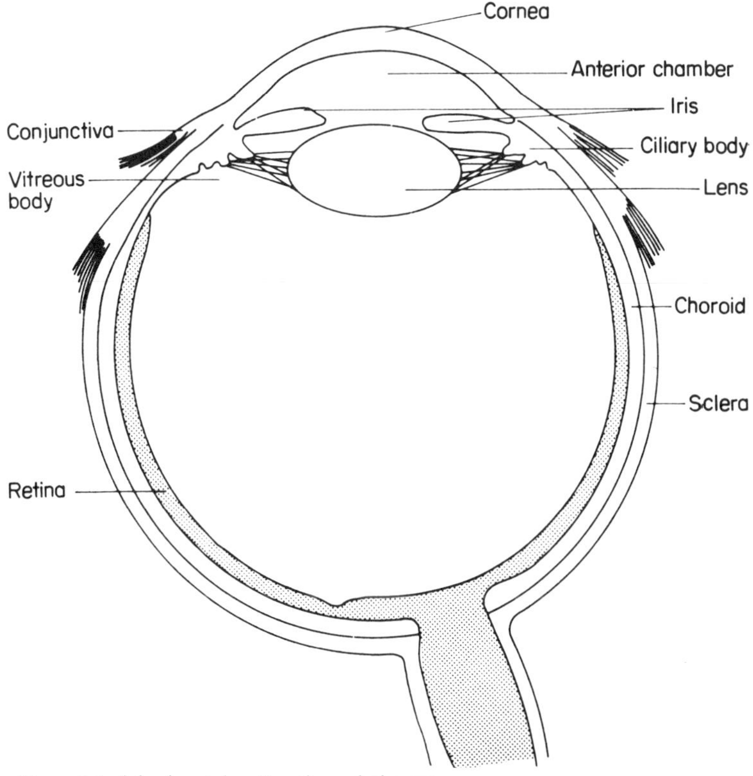

Figure 1.1. A horizontal section through the eye

The middle layer is mainly vascular and is made up of the choroid, ciliary body and iris. The innermost layer is the retina – predominantly nervous tissue. Within the three coats the eye is divided into two sections by the lens. The frontal section contains the aqueous humour, and is itself divided into anterior and posterior chambers by the iris. The section behind the lens contains the vitreous humour.

The important sites for drug action are the iris and ciliary muscle, the blood vessels, the extra-ocular muscles and the lacrimal gland (*Figure 1.2*). The iris and ciliary body are essentially composed of smooth muscle units which are innervated by the autonomic nervous system. The blood vessels and the levator

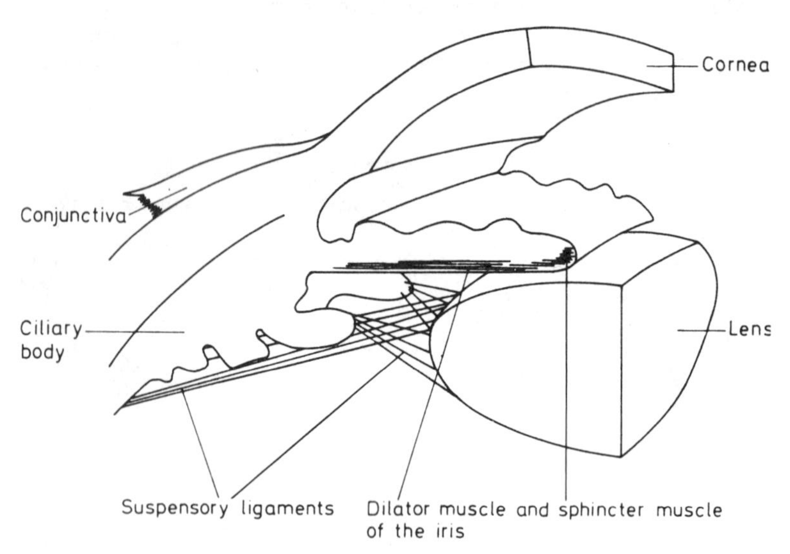

Figure 1.2. Schematic diagram of the anterior chamber

palpebrae also contain smooth muscle units. Other extra-ocular muscles (for example, the recti and obliques) consist of striated muscle innervated by the somatic nervous system. The lacrimal glands receive both parasympathetic and sympathetic innervation, although in the latter case the supply may be only to the blood vessels, not the secretory cells.

The iris

The iris is a forward extension of the choroid and arises from the anterior face of the ciliary body. The pupil forms a central

Preface to the first edition

This book is concerned with all aspects of drug action in the eye which are of relevance to the ophthalmic optician or optometrist. It contains a concise discussion of relevant ocular anatomy and physiology so that that actions and uses of drugs may be put on a rational basis. The major groups of drugs used by the ophthalmic optician are discussed in detail and each section includes a discussion of the practical aspects of drug use. The groups of drugs covered include: cycloplegics, mydriatics, miotics, local anaesthetics, staining agents, antimicrobial agents, decongestants, antihistamines and anti-inflammatory agents. There are separate chapters dealing with first aid and emergency measures and with the legal aspects of the sale and supply of drugs commonly used in the eye. At the time of going to press, some of the Acts covering the sale and supply of drugs are being repealed and replaced by new Acts. However, it is not expected that the changes will significantly modify the requirements placed on the ophthalmic optician.

A chapter on the ocular effects of drugs used systemically contains details not only of the ocular effects of named drugs, but also of the conditions for which these drugs are likely to be used. There is a section on the solutions used in contact lens work, which deals both with the drugs used by the ophthalmic optician during fitting and those used by the patient in lens care. In each chapter the official (BP or BPC) preparations are listed and there is a section on the formulation of preparations for the eye. The physicochemical properties underlying the absorption of drugs through the cornea are expanded so that the rationale behind the choice of a particular drug form and formulation may be understood.

As a whole, the book is designed to be a detailed yet concise account of drugs and their use in the eye.

It is intended that this book should be a working guide for qualified ophthalmic opticians/optometrists. However, since it contains considerable discussion not only of the practical aspects of drug use in the eye but also of the basic principles underlying this use, it should be invaluable to students reading for a degree or diploma in ophthalmic optics and optometry. As the book contains a detailed account of drug action in the eye and information on the diagnostic techniques, it should be of value to the medical and allied professions. In this context it should be of interest to students of medicine, physiology, pharmacy, and pharmacology and also to nurses.

M.J.V.
B.C.

aperture in the iris. Two different muscle layers are contained within the iris tissue. These are the dilator pupillae and the sphincter pupillae. The dilator pupillae is made up of modified pigment epithelial cells which are arranged radially. These cells receive sympathetic innervation. Stimulation of this nerve supply produces a contraction of the cells resulting in an increase in pupil size (mydriasis). The sphincter pupillae is a typical smooth muscle structure with its fibres arranged in a circular fashion. It receives parasympathetic innervation. An increase in the activity of this nerve supply produces a contraction of the cells resulting in a decrease in pupil size (miosis); a decrease in the activity of this nerve supply allows a relaxation of the muscle cells resulting in an increase in pupil size (mydriasis).

The ciliary muscle

The ciliary muscle is composed of smooth muscle units and is located against the inner surface of the anterior portion of the sclera. Most of the muscle fibres originate from the scleral region and run in meridional and radial groups. The innermost muscle fibres are arranged in the form of a sphincter and these receive parasympathetic innervation. Stimulation of the parasympathetic nerves causes contraction of the muscle, which in turn results in a lessening of the tension on the suspensory ligaments. This decrease in tension allows the lens to change shape, so that it becomes more convex. The change in the radius of curvature increases the power of the lens and allows near objects to be brought into focus at the retina.

Innervation of the iris and ciliary muscle

The subdivision of the nervous system into central and peripheral components is shown in *Figure 1.3*.

The peripheral component is further subdivided into afferent or sensory neurones, which carry impulses into the central nervous system, and efferent neurones, which carry impulses out of the central nervous system. Efferent neurones may be either somatic, that is, those supplying skeletal muscle, or autonomic. The ciliary muscle and iris are innervated by the autonomic division.

There are two branches of the autonomic nervous system known as parasympathetic and sympathetic. As can be seen in *Figures 1.4*

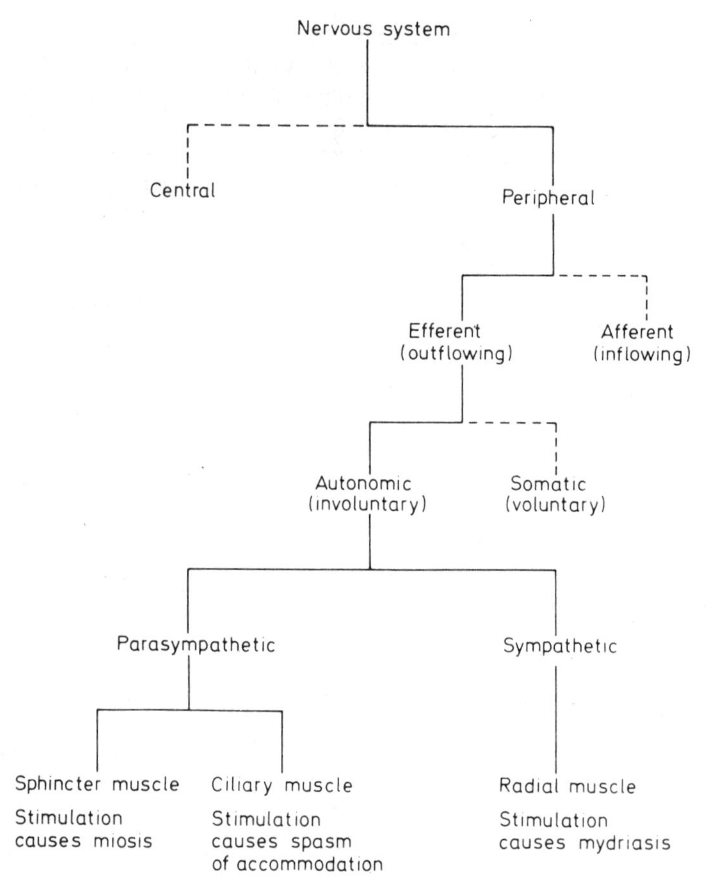

Figure 1.3. Subdivision of the nervous system on an anatomical basis showing autonomic innervation of the eye

and *1.5*, the autonomic nerves supply most of the body's organs and tissues.

The iris receives both parasympathetic and sympathetic innervation, the sphincter pupillae being innervated by the parasympathetic division and the dilator pupillae by the sympathetic. The parasympathetic nerve supply originates in the 3rd nerve nucleus in the central nervous system, which it leaves via the 3rd cranial nerve. There is a long preganglionic nerve which terminates in the ciliary ganglion (*Figure 1.6*). A short postganglionic fibre leaves the ganglion to innervate the sphincter pupillae. The lacrimal glands also receive parasympathetic nerves which

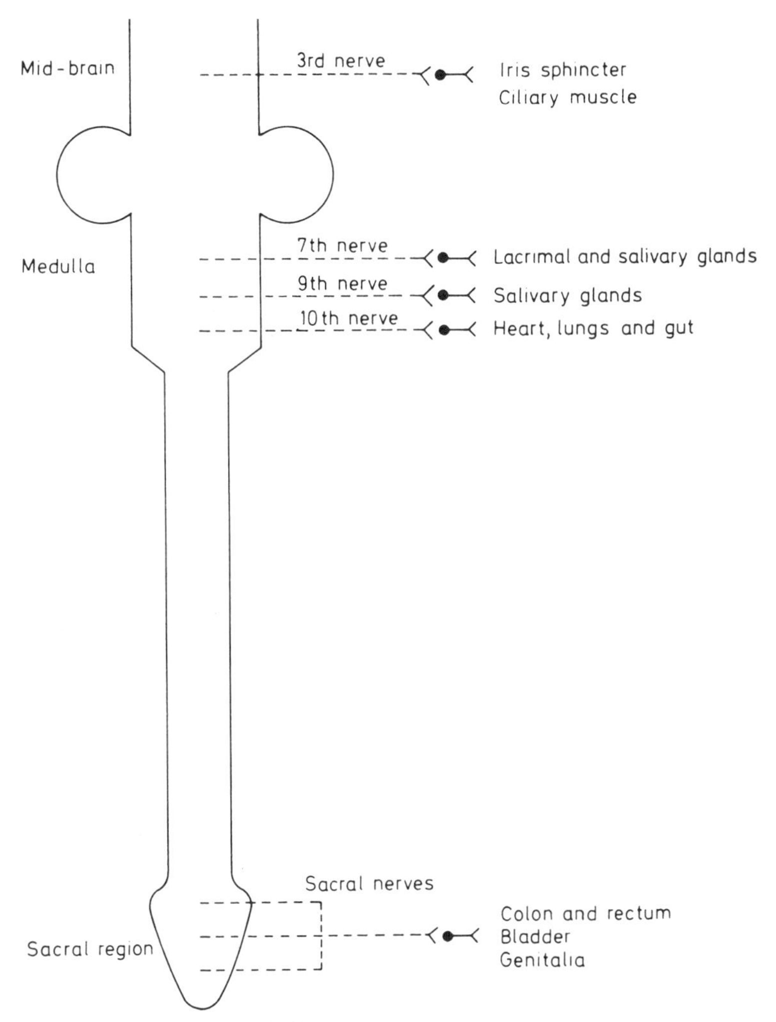

Figure 1.4. The parasympathetic nervous system

originate in the 7th nerve nucleus and pass to the glands via the sphenopalatine ganglion. The sympathetic nerve supply leaves the spinal cord from the cervical and upper thoracic segments and ascends to the superior cervical ganglion where it terminates. A long postganglionic fibres which leaves the superior cervical ganglion then innervates the dilator pupillae muscle. Sympathetic postganglionic fibres from the superior cervical ganglion also innervate the lacrimal glands and the smooth muscle of the lids.

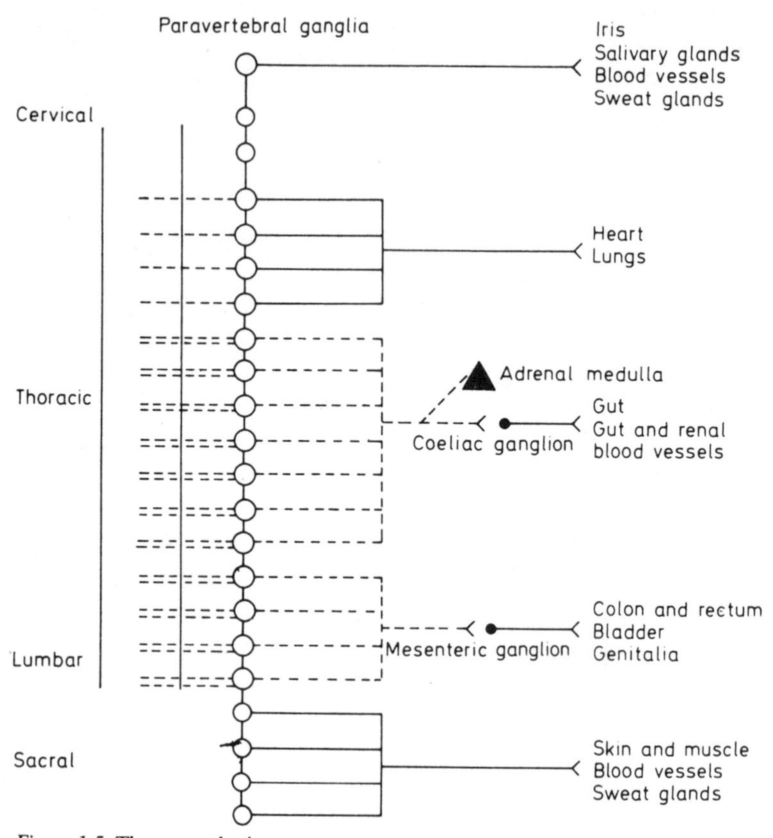

Figure 1.5. The sympathetic nervous system

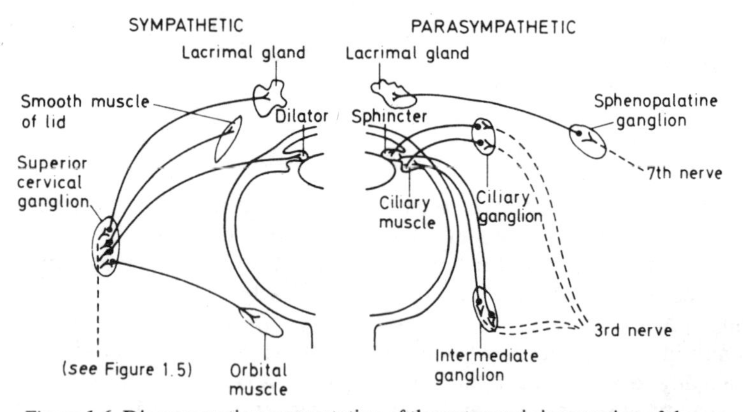

Figure 1.6. Diagrammatic representation of the autonomic innervation of the eye

The effective nervous supply to the ciliary muscle is part of the parasympathetic system following a pathway similar to the nervous supply of the sphincter pupillae.

Neurohumoural transmission and the receptor theory of drug action

It has been established in the previous section that the autonomic nervous system is composed of efferent pathways comprising two neurones, termed preganglionic and postganglionic, with a synapse between them. A synapse is the region of close approach between two neurones. At this site there is a small gap between the two which is called the synaptic cleft (20 nm). Thus, when a nerve impulse passes down a preganglionic fibre the impulse is not continuous in the postganglionic fibre. When the nerve impulse reaches the end of a preganglionic fibre, it causes a chemical substance to be released. This diffuses across the synaptic cleft and comes into close contact with specially sensitive sites on the postganglionic neurone. The result of this close contact is an alteration of the electrical properties of the postganglionic fibre and the initiation of a nerve impulse. A similar sequence of events occurs at the synapse between the postganglionic autonomic neurone and its effector cell (the neuroeffector junction) and also between a somatic motor neurone and a skeletal muscle cell (the neuromuscular junction). This process of chemical transmission of nerve impulses is known as neurohumoural transmission.

Two substances have been identified as chemical transmitters of peripheral nervous activity, they are acetylcholine and noradrenaline. Acetylcholine is the transmitter liberated from presynaptic nerves to act upon postsynaptic effector cells at the following four sites:

(1) All ganglia.
(2) At postganglionic parasympathetic nerve endings.
(3) At the skeletal neuromuscular junction.
(4) At a few postganglionic sympathetic nerve endings, that is, postganglionic sympathetic fibres concerned with sweating and with the dilatation of blood vessels of skeletal muscle.

Noradrenaline is a transmitter at postganglionic sympathetic nerve endings with the exception of the two areas listed under (4) above. Historically, the transmitter was once thought to be adrenaline and these nerves are still referred to as adrenergic.

The following sequence of events takes place at all synapses:

(1) The transmitter is synthesized and stored in the nerve ending.
(2) Nerve impulses cause the release of the transmitter into the synaptic cleft.
(3) The transmitter diffuses across the gap and activates the sensitive postsynaptic area (receptor).
(4) The transmitter is rendered inactive.

Sites receiving cholinergic innervation

A diagrammatic representation of the sequence of events occurring at such a site is shown in *Figure 1.7.*

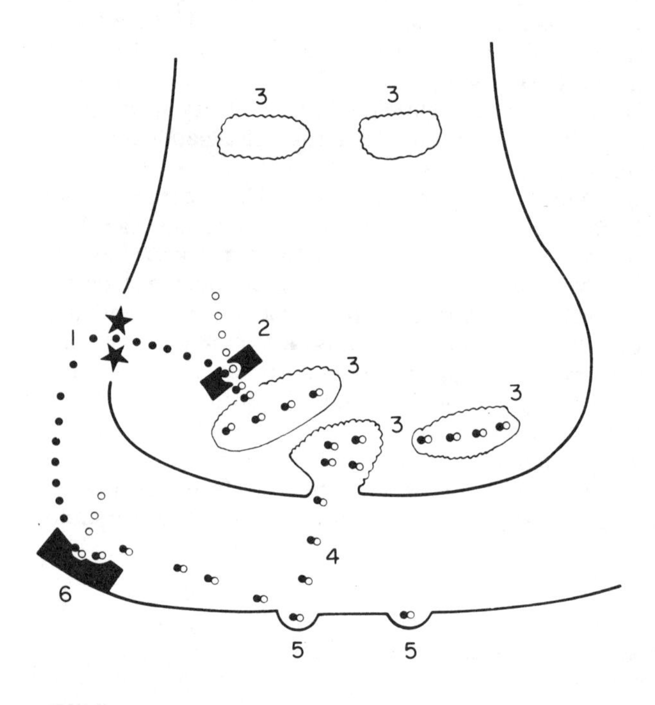

Figure 1.7. Diagrammatic representation of a cholinergic synapse showing synthesis, release and inactivation of acetylcholine (●○)

The chemical substance choline is concentrated within the presynaptic nerve ending by means of a specialized transport system known as the choline pump (1). This transport system requires energy and is susceptible to metabolic poisons. Once inside the nerve ending, the choline combines with acetyl coenzyme A to form the ester acetylcholine (*Figure 1.7*). This reaction is catalysed by the enzyme choline acetylase (2).

The acetylcholine so formed is stored in specially differentiated parts of the nerve ending called vesicles (3). When a nerve impulse passes down the presynaptic neurone vesicular acetylcholine is released into the synaptic cleft (4) and diffuses to the postsynaptic site (5). After activating the receptor the acetylcholine diffuses away and comes into contact with an enzyme, acetylcholinesterase (6). This enzyme is found predominantly in the synaptic regions and causes a reversal of the process depicted in (2). The choline so liberated is available for re-uptake into the presynaptic nerve for the synthesis of new transmitter material.

Drugs may interfere at various points in this sequence of events. The stages subject to drug interference and the types of drug active at these stages are shown in *Figure 1.8*.

Drugs such as hemicholinium will compete with choline for the choline pump (1), so that choline cannot gain access into the presynaptic nerve. This prevents synthesis of new transmitter material and eventual loss of transmission. Other drugs such as triethylcholine are taken up by the choline pump and compete with choline inside the nerve, resulting in the production of acetyltriethylcholine (2), which is stored in a similar way to acetylcholine (3). However, acetyltriethylcholine is relatively ineffective after release (4).

All the drugs mentioned above will act at any synapse at which acetylcholine is the transmitter and, while they may be of value and interest to the pharmacologist or physiologist, they are not of therapeutic value.

The next stage in neurotransmission is stage (5), the process of transmitter receptor interaction. The drugs which act at this point in the sequence are of therapeutic value. Drugs acting like acetylcholine are called agonists (for example, pilocarpine), and those which prevent the action of acetylcholine by blocking access to the receptor are known as antagonists (for example, atropine). The final stage in the sequence of events occurring at these synapses is the enzymatic breakdown of acetylcholine and this stage is also susceptible to drug interference. Thus some drugs combine with the enzyme acetylcholinesterase (6) which is present in the synaptic cleft. These drugs, the anticholinesterases, interact with the active sites on the acetylcholinesterase, but are not hydrolysed. They therefore prevent the access of acetylcholine to the enzyme and cause an increased concentration of acetylcholine in the synaptic cleft, thereby increasing the number of transmitter receptor interactions. Another family of cholinesterase enzymes also occurs in the body associated with serum and with muscle.

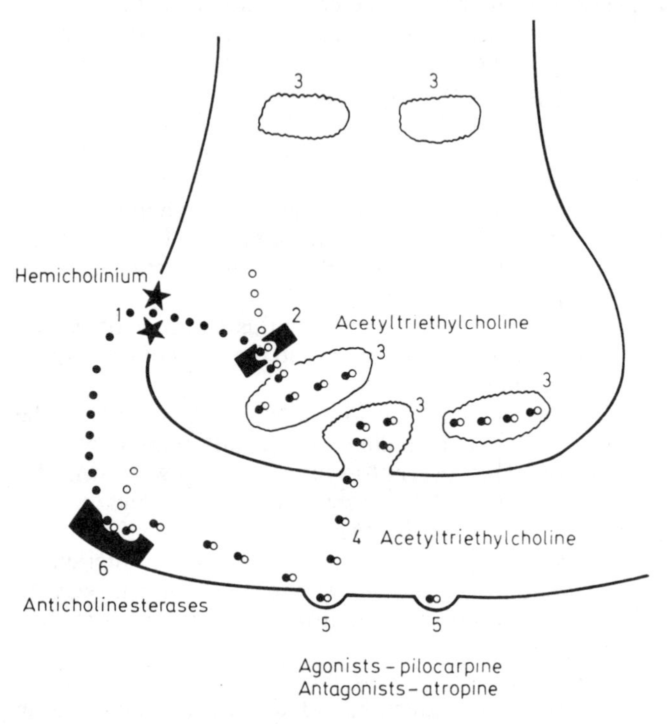

Figure 1.8. Diagrammatic representation of the sites at which drugs may interfere with the functioning of a cholinergic synapse

These enzymes, often called pseudocholinesterases, will hydrolyse other esters and some amides as well as acetylcholine, and are also susceptible to attack by the anticholinesterases. The function of this group is probably to hydrolyse esters which either enter the body in the food, or occur as the result of metabolic processes.

In order to fully understand the interactions of drugs at receptor sites a more detailed description of the transmitter receptor interaction is required. It is postulated that the interaction between acetylcholine and the receptor relies on the physico-chemical properties of the transmitter substance and the receptor site. Thus, acetylcholine in solution can be regarded as a carbon skeleton, which is neutral, carrying chemical substituents which bear small electrical charges. As shown in *Figure 1.9* a small positive charge resides on the nitrogen and small negative charges on the oxygen substituents. The receptor similarly will carry small electrical charges distributed in such a way that the negative charges of the transmitter correspond to the positive charges of the receptor and the positive charges of the transmitter correspond to the negative charges of the receptor. When these two come into

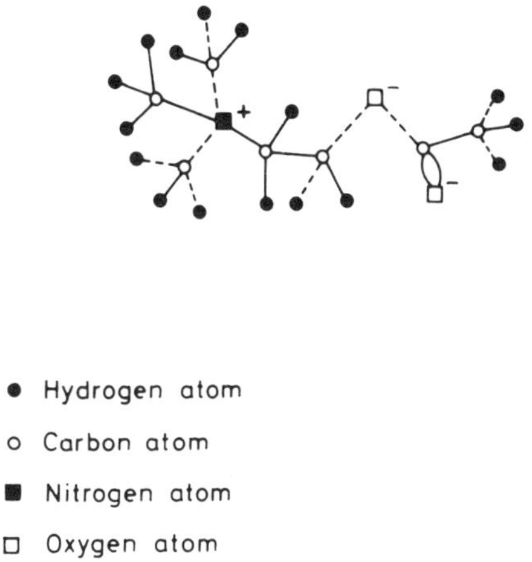

● Hydrogen atom

○ Carbon atom

■ Nitrogen atom

□ Oxygen atom

Figure 1.9. Chemical structure of the acetylcholine molecule to show the probable distribution of the electropositive (+) and electronegative (−) charges

contact the resulting interaction causes changes in the structure of the receptor. It is these changes which initiate a sequence of events such as nerve action potential or muscle contraction. Acetylcholine is not a rigid structure and may exist in solution in many conformations. While one particular conformation may be satisfactory for interaction with, say, a receptor on a nerve (that is, at a ganglion), this conformation will not necessarily be suitable for interaction with the receptor on a skeletal muscle or on an autonomic effector cell. Therefore, while acetylcholine is the transmitter at all these sites, it is postulated that the receptors differ. Because of this fact it is possible to select drugs which have a specific action on one type of receptor. The prime example of such a drug is the alkaloid muscarine. Thus, this drug has a similar configuration to that of acetylcholine when acetylcholine interacts with the autonomic effector cells innervated by postganglionic parasympathetic nerves. It is therefore selective for the receptors on these cells and has led to their description as muscarinic receptors. Muscarine does not have the same configuration as acetylcholine when this latter substance acts in the ganglia or on a skeletal muscle receptor. Another alkaloid, nicotine, has some structural similarity to the conformation of acetylcholine when it acts at these other two sites. Thus, these receptors are classified as nicotinic receptors.

Drug action

At the muscarinic receptor

This may be subdivided into drugs which interact with the receptor to produce an effect similar to acetylcholine and those which interact with the receptor but do not produce those changes which are associated with an acetylcholine interaction. However, by virtue of the drug occupation of receptor sites, access of acetylcholine is prevented and transmission at the site is blocked. Drugs acting like acetylcholine at the receptor are called muscarinic agonists, for example, pilocarpine; drugs blocking the action of acetylcholine at the receptor are called muscarinic antagonists, for example, atropine, homatropine and cyclopentolate. Thus, the muscarinic agonists will act on the ciliary muscle and sphincter pupillae to cause spasm of accommodation and miosis. Muscarinic antagonists will act at these sites to produce paralysis of accommodation (cycloplegia) and mydriasis.

At the nicotinic receptors

Receptors for acetylcholine are similar in all ganglia whether sympathetic or parasympathetic. However, the nicotinic receptor at the ganglion is not identical with the nicotinic receptor at the neuromuscular junction. The evidence for this statement is based on the fact that some drugs will interact with the ganglion receptors to prevent the action of acetylcholine but are relatively ineffective at the receptors on skeletal muscle. Hexamethonium and pempidine are efficient ganglion blocking agents but do not cause muscle paralysis. Drugs such as tubocurarine and pancuronium are potent antagonists at skeletal muscle receptors, causing muscle paralysis, but they are relatively ineffective in blocking the ganglion receptors. The effects of ganglion blockade are complex because many tissues are innervated by both sympathetic and parasympathetic nerves and therefore the end result depends on the relative importance of the two systems. Since the parasympathetic system is dominant in the iris and ciliary muscle, the ocular effects of ganglion blockade are mydriasis and cycloplegia. Drugs with blocking actions at the nicotinic receptors in skeletal muscle are not used by the ophthalmic optician but are reserved primarily to produce muscular relaxation in surgery or electroconvulsive therapy.

Sites receiving adrenergic innervation

A diagrammatic representation of the sequence of events at an adrenergic synapse is shown in *Figure 1.10.*

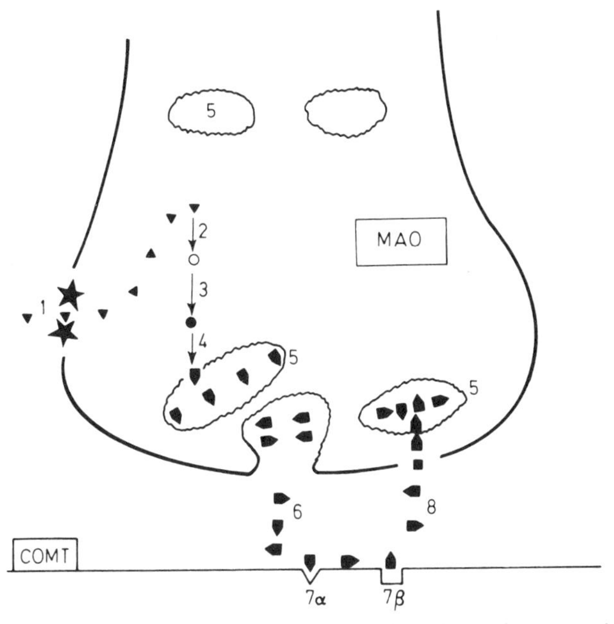

Figure 1.10. Diagrammatic representation of an adrenergic synapse showing the synthesis, release and inactivation of noradrenaline (●)

The amino acid tyrosine (1) is taken up by the presynaptic nerve. Once inside the nerve a sequence of enzymatic steps takes place (2–4) leading to the formation of noradrenaline which is concentrated in granulated vesicles (5) present in the nerve endings. When a nerve impulse reaches the terminal the stored noradrenaline is released into the synaptic cleft (6) and diffuses to the postsynaptic receptor site (7, α or β). After interaction with the receptor site (often called the adrenoreceptor), the noradrenaline diffuses back across the synaptic cleft and is actively taken up into the presynaptic neurone (8). This uptake process is energy dependent, saturable and relatively specific. The number of transmitter–receptor interactions is related to the concentration of the transmitter in the area of the receptor. Thus, when noradrenaline is removed from the synaptic cleft and concentrated within the nerve, the number of transmitter–receptor interactions declines and the transmitter effect is reduced and eventually abolished. This mechanism is the major one involved in the inactivation of noradrenaline.

There are two enzymes in the body capable of metabolizing noradrenaline: monoamine oxidase (MAO) and catechol O-methyl transferase (COMT). However, these are not important for the inactivation of noradrenaline released from the presynaptic nerve. Large quantities of monoamine oxidase are found in the liver where its function is to metabolize naturally occurring amines present in the blood stream. It is also present in nerves and it oxidizes a proportion of the intraneuronal noradrenaline which is not in the vesicle. Catechol O-methyl transferase occurs mainly within the effector cells and in the liver and kidney. It is an enzyme of low specificity and has other substrates.

The points at which drugs may interfere with the sequence are shown in *Figure 1.11*.

The various steps in the synthetic pathway of noradrenaline are subject to inhibition leading to a decreased production of the natural transmitter and loss of transmission. Alpha-methyldopa (1) is a compound capable of entering the synthetic pathway. Alpha-methylnoradrenaline is produced rather than noradrenaline (2–4). The alphamethyl compound is stored (5) and released in the same way as noradrenaline. Release of noradrenaline following a nerve impulse may be blocked by a group of drugs known as adrenergic neurone blocking agents (6). These compounds, which include guanethidine, bethanidine and bretylium, appear to concentrate in the fine endings of postganglionic sympathetic nerves and thereby interfere with the release process. Alpha-methyldopa and the adrenergic neurone blocking agents are used to reduce sympathetic activity in an attempt to combat hypertension. Other drugs act postsynaptically interfering with transmitter–receptor interactions. There are at least two types of receptor found at the postganglionic sympathetic nerve effector cell junction. These two receptors are designated α and β (7). In general, interaction between the transmitter and an α-receptor causes an excitatory response (for example, contraction of the smooth muscle cells of the dilator pupillae) and interaction between the transmitter and a β-receptor causes an inhibitory response. The exception to this is that interaction with the β-receptors in the heart results in an increased rate and force of contraction. It is now recognized that there are two subdivisions of β-receptors: those in cardiac tissue being designated β_1 and those in smooth muscle (such as in the respiratory tract) β_2.

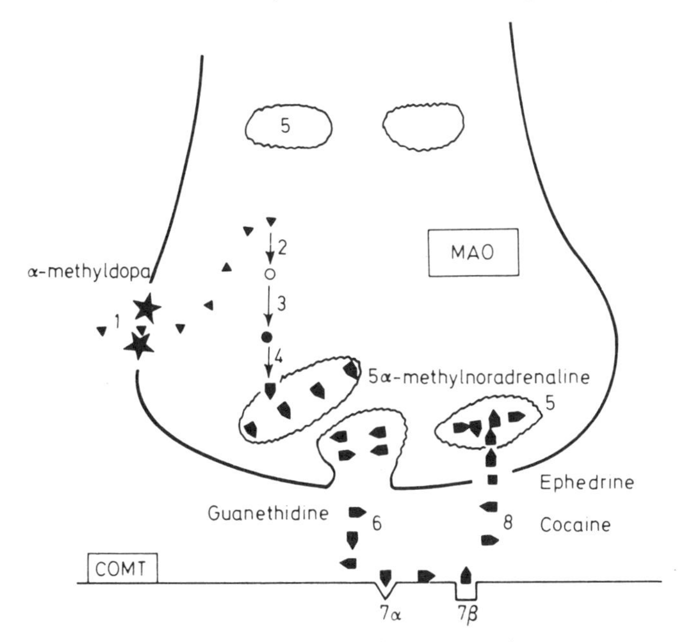

Figure 1.11. Diagrammatic representation of the sites at which drugs may interfere with the functioning of an adrenergic synapse

Noradrenaline is predominantly active at the α-receptors. Adrenaline, the hormone synthesized and stored in the adrenal medulla and released into the blood stream in response to stress, has approximately equal activity at both types of receptors. There are a large number of synthetic compounds which have sufficient structural resemblance to the natural transmitter to be able to interact with either of the receptor types and produce a response. Among these compounds with agonist activity are phenylephrine (7 α) which acts almost exclusively with α-adrenoceptors and isoprenaline (7 β) which acts almost exclusively with β-adrenoceptors. Phenylephrine and isoprenaline are known as directly acting sympathomimetics. Certain compounds, for example ephedrine, have some direct action but the major part of their activity is indirect. Thus, they penetrate the neurone (8), enter the vesicular store and cause noradrenaline to be released (6) into the synaptic cleft. The released noradrenaline is available for interaction with the receptor. As was stated earlier, the most important factor in decreasing transmitter–receptor interactions is the active re-uptake of noradrenaline into the nerve endings. A number of drugs, including cocaine, are able to block this active process (8) and thus maintain a higher concentration of noradrenaline in the vicinity of the receptor. Use of either type of indirectly acting drug will result in increased sympathetic activity, for example, contraction of the dilator pupillae.

Some drugs can inhibit the interactions of transmitters and receptors. These drugs bear sufficient structural resemblance to noradrenaline to interact with the receptors but after combination with the receptor produce no response (antagonists). The presence of the antagonist means that the number of transmitter–receptor interactions is reduced because noradrenaline cannot gain access to the receptor. These antagonists at the adrenoceptors are divided into two types depending on whether they block either α-receptors (phenoxybenzamine, phentolamine, 7α) or β-receptors (propranolol, 7β). Thymoxamine, an α-adrenoceptor antagonist, is now available in an ophthalmic preparation for the reversal of mydriasis induced by α-agonists (for example, phenylephrine); timolol, a β-adrenoceptor antagonist is used to lower intraocular pressure in open angle glaucoma.

2

Factors affecting drug absorption

The drugs used by the ophthalmic optician are usually applied to the eye in the form of drops or, more rarely, ointments. Topical application is used in order to restrict the site of drug action to the eye, reduce the possibility of unwanted effects and reduce the quantity of drug used. However, in order to achieve this localized action, there must be sufficient absorption from the site of application to give and maintain effective drug concentrations at the site of action.

The absorptive surface in question is the cornea. Any drug absorbed by the conjunctiva enters the systemic circulation and is lost from the eye. Drug loss may also occur through the puncta, unless pressure is placed on the nasal canthus. The cornea consists essentially of three parts: the epithelium, the stroma and the endothelium. The epithelium and the endothelium both have a high lipid content and are therefore readily penetrated by compounds with high lipid solubility (that is, non-polar, un-ionized structures). The stroma lying between the epithelium and endothelium is an aqueous structure containing some 75–80% water. This layer will therefore be more readily penetrated by compounds which are polar and water-soluble (for example, ionized particles). Thus, in order for a drug placed on the cornea to penetrate all three parts of the cornea, it must be both lipid and water soluble. Hydrocarbon structures confer lipid solubility on a compound, water solubility depends on the presence of hydrophilic or polar moieties such as hydroxyl (OH) or ionized groups in the structure. A structure containing only hydrocarbon groupings will exhibit high lipid solubility and poor water solubility. A compound having polar substituents will show high water solubility and poor lipid solubility. In either of these extreme cases the compound would not be capable of crossing all the layers of the cornea. A compound whose structure contains both some non-polar and some polar groupings will show both lipid and water solubility and will therefore be able to cross the cornea.

The majority of drugs used by the optician are weak bases. A weak base is used as its water-soluble acid salt, for example, phenylephrine hydrochloride (*Figure 2.1a*). Thus, in solution this will exist in both the ionized (water-soluble) and un-ionized (lipid-soluble) forms in equilibrium according to *Figure 2.1(b)* and satisfy the above criteria for corneal penetration. (For a more detailed discussion of the factors affecting the absorption of a weak base *see* Appendix.)

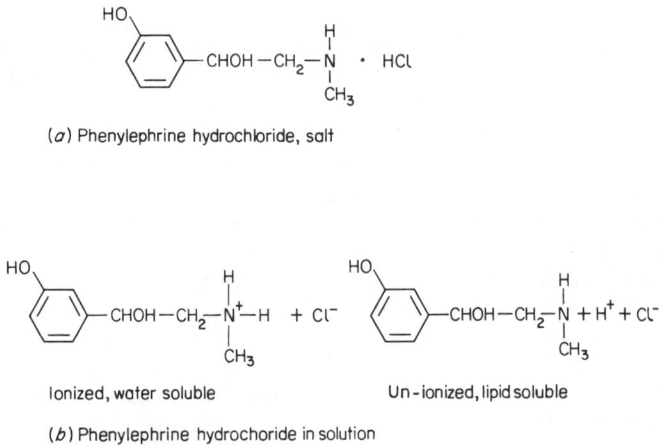

(*a*) Phenylephrine hydrochloride, salt

Ionized, water soluble

Un-ionized, lipid soluble

(*b*) Phenylephrine hydrochoride in solution

Figure 2.1. Dissociation of phenylephrine hydrochloride (a weak base) in aqueous solution

Some synthetic drugs are quaternary amines and highly ionized in solution at any pH. Thus, they have poor lipid solubility and poor penetration of the cornea (for example, neostigmine, carbachol). Surface active agents (such as benzalkonium chloride) which reduce surface tension are able to increase the permeability of the cornea and aid penetration by drugs. Benzalkonium chloride may be used for this purpose with carbachol (a muscarinic agonist) or neostigmine (an anticholinesterase).

As the absorption is a passive process, an important factor for absorption is the concentration gradient. In passive absorption drugs diffuse down a concentration gradient from an area of high concentration. Solutions used by opticians are relatively concentrated. The concentration in the aqueous humour or tissues of the eye does not approach that in the solution and provides no hindrance to absorption.

Eye drops have often been formulated to be isotonic but it seems that variations from isotonicity do no great harm to the eye. Hypertonicity may cause stinging and hypotonicity may increase the permeability of the cornea.

Another factor which may influence the absorption of drugs is damage to the corneal tissues. Removal or abrasion of the epithelium can increase the absorption of drugs with low lipid solubility (although the endothelium must still be crossed). This may lead to unexpectedly high concentrations of drugs in the eye.

One final consideration is the form in which the drug is presented to the eye: aqueous drops, oily drops or ointments. Aqueous drops are most commonly used. These are convenient, but have the disadvantage of a short contact time with the cornea and a short time for absorption. Rapid initial absorption is therefore important and it would be preferable for the drug to be as little ionized as possible. Unfortunately, as most of the drugs used are the acid salts of weak bases, the major portion is in the ionized form. As drops are often placed in the lower fornix, a portion of the drug may be absorbed by the conjunctiva rather than the cornea and lost in the systemic circulation. Alternatively, some drug may be lost by passage through the puncta into the nose.

If oily drops or ointments are used, the base is more likely to be in its un-ionized lipid-soluble form suitable for absorption across the epithelium. This, together with the longer contact time of an oily preparation, would suggest that absorption would be more rapid, provided that the base was capable of ionizing in the aqueous stroma. One possible disadvantage is that the drug may be more soluble in the oily base of the ointment than in the lipid layer of the cornea and therefore be trapped in the ointment.

3

Cycloplegics

A cycloplegic drug causes paralysis of accommodation. It thus renders the eye unable to focus on near objects and does this by inhibiting the effect of acetylcholine released from postganglionic parasympathetic nerves. Atropine is a typical example of this class of drug and, while it may not now be so widely used, it still serves as a model with which to compare other available cycloplegic drugs. Atropine competes with acetylcholine for the receptor sites on smooth muscle, in this case on the ciliary body. The action of acetylcholine is thus prevented, the ciliary muscle is not responsive to parasympathetic nerve activity and relaxes. The tension on the suspensory ligaments of the lens is increased so that the lens itself becomes less convex and accommodation is inhibited.

Atropine and atropine-like drugs do not affect nerve impulses, neither do they prevent the release of acetylcholine.

As the sphincter pupillae also has postganglionic parasympathetic innervation, the constrictor action of acetylcholine at this site is also antagonized and the pupil dilates. With this group of drugs cycloplegia is always accompanied by mydriasis.

Atropine

Atropine is a naturally occurring alkaloid extracted from various solanaceous plants including *Atropa belladonna* (deadly nightshade) and *Datura stramonium* (thorn apple). It is an organic ester formed by the combination of an aromatic acid and an organic base tropine (*Figure 3.1*).

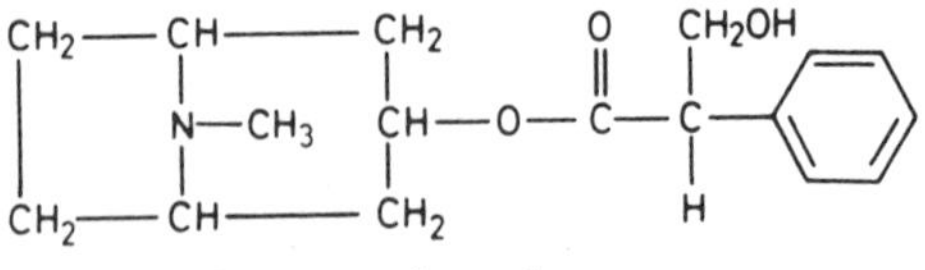

```
CH2——CH———CH2          O   CH2OH
|     |      |          ||    |
|     N—CH3  CH—O—C—C—(C6H5)
|     |      |                |
CH2——CH———CH2                 H
```

Figure 3.1. Structure of atropine

Mode of action

Atropine is a competitive antagonist of acetylcholine. That is to say, the drug atropine and the transmitter acetylcholine are in direct competition with each other for the acetylcholine muscarinic receptor sites on smooth muscle. If acetylcholine occupies the receptor then a sequence of events occurs resulting in muscle contraction. If atropine occupies the receptor this is not available to acetylcholine, the sequence is prevented and the muscle is relaxed. Two factors are important in determining the number of receptors occupied by a drug. First, the affinity of the drug for the receptor and second, the concentration of the drug in the tissue fluid surrounding the receptor (the biophase). Atropine has a high affinity for the muscarinic receptor and therefore is effective in relatively low concentrations however, and its effects are long lived and difficult to reverse.

Actions in the eye

Sphincter pupillae

After instillation of a 1% solution of atropine sulphate mydriasis usually commences within 10–15 minutes and the maximum effect is reached within 30–40 minutes. At this time the pupil is widely dilated and the response to light is abolished. The return to normal pupil size is slow and may take several (up to 10) days.

Ciliary muscle

As the ciliary muscle lies deeper in the eye than the sphincter pupillae, the diffusion path of the drug is longer. The onset of cycloplegia is thus later than the onset of mydriasis. A slight decrease in accommodation may be noted within 30 minutes. The decrease in amplitude progresses slowly, a maximum effect occurring in 1–3 hours. Some recovery from the cycloplegic effect of atropine begins in 2–3 days. The ability to read fine print has usually returned in 3 days but recovery of full amplitude of accommodation may take up to 10 days.

Secretions

Atropine reduces the secretory activity of the salivary, gastro-intestinal, lacrimal and sweat glands. Its effect on the lacrimal gland is of interest because it alters the character of the secretion, reducing its water content and thereby increasing the concentration of solutes. Tear fluid contains an enzyme, lysosyme, which is active against a number of bacteria by action on their cell walls. Atropine will increase the concentration of lysosyme in lacrimal fluid, thus increasing its antibacterial properties.

Intraocular pressure

There is a possibility that the use of atropine may lead to an increase in intraocular pressure. Dilation of the pupil results in the iris being thickened and 'piled up' in the angle of the anterior chamber. In the case of a shallow anterior chamber the outflow of fluid through the trabecular meshwork, the canal of Schlemm and the aqueous veins may be hindered. The alternate contractions and relaxations of the ciliary muscle during normal accommodation exert a varying pull on the scleral spur and have some effect in changing the size of, and pressure within, the canal. If pressure increases fluid passes more easily into the intrascleral veins than back into the anterior chamber. If the pressure falls in the canal, fluid will pass into the canal from the anterior chamber but will not be drawn back from the intrascleral veins, which have valves to prevent this happening. Thus, normal movements of the ciliary muscle during accommodation produce a pumping action aiding outflow of fluid from the eye. Atropine will prevent this action, again tending to reduce outflow. Other factors which have been postulated to contribute to the increase in intraocular pressure after atropine are:

(1) Facilitated inflow and impeded outflow of blood due to the immobilization of the ciliary muscle. This may result in intraocular congestion.
(2) An increase in capillary permeability possibly by a mechanism involving histamine release.

The above considerations pose the question: Why is it that atropine does not always produce an increase in intraocular pressure? The answer lies in the efficient regulatory mechanisms which the normal eye can bring into play. As pressure rises the outflow of aqueous increases and the rate of aqueous secretion decreases. This will tend to keep intraocular pressure constant. However, with increased age, changes take place in the eye which reduce the efficiency of the regulatory mechanisms.

(1) The sclera becomes less distensible by losing its elastic properties. Any increase in intraocular congestion will thus produce a relatively greater effect on the pressure.
(2) Bundles of fibres at the filtration angle become sclerosed and thickened, thus impeding outflow.

(3) Pigment from the iris and the ciliary body becomes trapped and collects between the fibres impeding outflow.
(4) The lens increases in size, both in thickness and in diameter. The space between the periphery of the lens and the ciliary processes is decreased, and the iris is pushed forward producing a shallow anterior chamber and a narrowing of the angle.

In many people these changes do not exceed certain critical limits and therefore there is no increase in intraocular pressure. In other people the changes occur to an unusual degree, particularly in the hyperopic eye (small globe, normal lens and large ciliary muscle). In these people there is a tendency towards an increased intraocular pressure and under the stresses induced by atropine treatment an attack of glaucoma can ensue.

Thus, the use of atropine in older people can be very dangerous and, as it would be of little or no benefit in the refraction of this age group, its use should be rigorously avoided.

Other actions of atropine

The pharmacological basis of the use of atropine as a cycloplegic and mydriatic has been discussed. The parasympathetic nervous system innervates a wide variety of tissues. A blocking drug such as atropine will thus bring about a number of other actions if sufficient is absorbed following topical application, or if the drug is ingested.

The heart

Atropine will produce an increase in heart rate. The heart rate is determined by two separate factors: a sympathetic innervation which acts to produce an increase and a parasympathetic innervation (the vagus nerve) which causes a decrease. Atropine will block the effect of acetylcholine released from the vagus nerve resulting in a decrease in vagal tone and an increase in heart rate. This increase after atropine is usually more pronounced in children who have a higher vagal tone.

The circulation

In usual therapeutic doses atropine has little effect on the blood pressure, as most vascular tissue is without parasympathetic innervation. High doses of atropine may cause a fall in blood pressure but this is most likely mediated by an action on the central nervous system. Atropine may also cause dilatation in cutaneous blood vessels which is most prominent in the blush area. It is thought likely that this effect is due to release of histamine.

Smooth muscle

Atropine will reduce the tone and motility of most smooth muscle tissue, although some tissues with parasympathetic innervation (for example, the bladder) are more resistant to atropine than others. Both the tone and motility of the gastro-intestinal tract are decreased as is the tone and motility in the ureters. The smooth muscles of the bronchioles are also relaxed giving a slight decrease in airway resistance.

Secretions

Atropine will produce a reduction in the secretions of a number of glands with postganglionic parasympathetic innervation, for example, salivary glands, mucous secreting glands of the bronchioles and gastrointestinal tract. Secretion of gastric acid is less affected by atropine, as this is mainly under hormonal control. Atropine also reduces sweating. The eccrine sweat glands of the skin are innervated by postganglionic sympathetic nerve fibres. These fibres differ from other postganglionic sympathetic nerves in that the transmitter released is acetylcholine not noradrenaline.

Central effects

In very low doses atropine stimulates the central nervous system. Stimulation of the medulla and higher cerebral centres results in stimulation of respiration and slowing of the heart due to an action on vagal centres. This occurs before the concentration of atropine in peripheral tissues reaches a sufficient level to antagonize the acetylcholine released from postganglionic cholinergic nerves. Some slight agitation and excitement may be observed.

If the dose of atropine is progressively increased these effects become more prominent, developing through restlessness, irritability, disorientation and hallucinations to delirium. At very high doses the effect of atropine on the central nervous system changes to a depressant action and death ensues due to respiratory paralysis.

Atropine as a poison

Atropine poisoning due to absorption of the drug from ocular membranes or after passage through the puncta and subsequent absorption is a possibility. There have been some reports of psychotic changes and alterations in heart rate after the use of 1% atropine sulphate eye drops. These cases have led to the suggestion that if drops are to be used in children under the age of 7 years the concentration should not exceed 0.5%. The main danger from atropine is likely to be the accidental ingestion of the preparation, the ointment most often being supplied for use at home under the supervision of the parent.

The symptoms and signs of atropine poisoning are very typical, being a mixture of central and peripheral actions. They include dry mouth (causing difficulty in swallowing and speaking), thirst, nausea and vomiting, rapid respiration, dilated pupils, fixed blurred vision, photophobia, a hot dry flushed skin, high temperature and a rapid pulse. The patient becomes very restless and excited, he may become confused and suffer hallucinations. Motor incoordination ensues and the central excitation changes to central depression. Respiration becomes inadequate and ultimately leads to coma and death from respiratory depression.

It is perhaps a sobering thought that deaths in children have resulted from as little as 10 mg of atropine although in some cases there has been recovery from as much as 1 g. A tube of eye ointment containing 3 g of a 1% preparation would contain 30 mg of atropine sulphate.

Atropine irritation

This is a local reaction of the tissues shown only in a few patients who are hypersensitive to the drug. It is an allergic response requiring prior sensitization. Thus the response does not appear the first time the drug is used but only on subsequent occasions. The first step in the sensitization process is the combination of atropine with some substance which occurs naturally in the body. This substance, which is probably a protein, has antigenic properties. The antigen then induces the formation of antibodies (similar to those produced in response to an infection) by the plasma cells. The antibodies become attached to special cells in the body called mast cells. The mast cells are widely distributed throughout the body and are basophilic granular cells found in loose connective tissues and the peripheral blood. They have a high histamine, heparin and 5-hydroxytryptamine content. An allergic response requires antibody–antigen interaction. On the first exposure to atropine the antibody is produced. However, during the formation of the antibody, the concentration of atropine in the body is already declining due to metabolism and excretion. Therefore the relative concentrations of antibody and antigen are not optimal for interaction and no allergic response results.

On a second application of the drug the antibody levels rise much more rapidly and an interaction does take place. The reaction is thought to take place at the mast cell surface where the antibody is bound. An intracellular enzyme is activated which in turn causes the release of histamine, 5-hydroxytryptamine, various kinins and a substance previously known as 'slow reacting substance' (SRS), which is now believed to be a product of arachidonic acid metabolism. Although all these substances have been implicated in the allergic response, histamine appears to be the major active component. Histamine causes capillary dilatation and increased capillary permeability. This therefore leads to a warming of the skin and an area of oedema (collection of fluid within the tissue but outside the vascular system). A second action of histamine mediated through local nerve reflexes is to cause arteriolar vasodilatation. The affected area and surrounding tissue becomes red and hot. Itching is added to the already considerable discomfort. Thus, if atropine irritation were to occur after a second or later instillation, the skin round the eye would appear red and inflamed with swelling of the lids and conjunctiva. The conjunctiva would also appear red and inflamed and the patient would complain of itching and irritation. The condition frequently spreads to the upper nasal mucous membranes.

Although this condition is relatively rare and often only occurs after prolonged treatment with atropine (as in iritis), it is a condition of which the optician should be aware. The immediate and essential step when the allergic response occurs is withdrawal of the drug. If continued use of a cycloplegic is essential then some other compound must be substituted, preferably one with a chemical structure which is not closely related to the atropine molecule (*Figure 3.2*). Thus, if homatropine or hyoscine is substituted, it is likely that the patient will also be sensitive to these drugs since they have similar chemical structures. If cyclopentolate, which bears little chemical resemblance to atropine, is substituted there is less chance of cross-hypersensitivity.

After withdrawing the atropine the optician may refer the patient to a general practitioner for treatment. The patient should be assured that this is a temporary condition which will clear quickly. He should also be informed that this condition will recur if he is exposed to atropine-like compounds in the future. The doctor may provide some relief by the sparing use of adrenaline 1:1000,

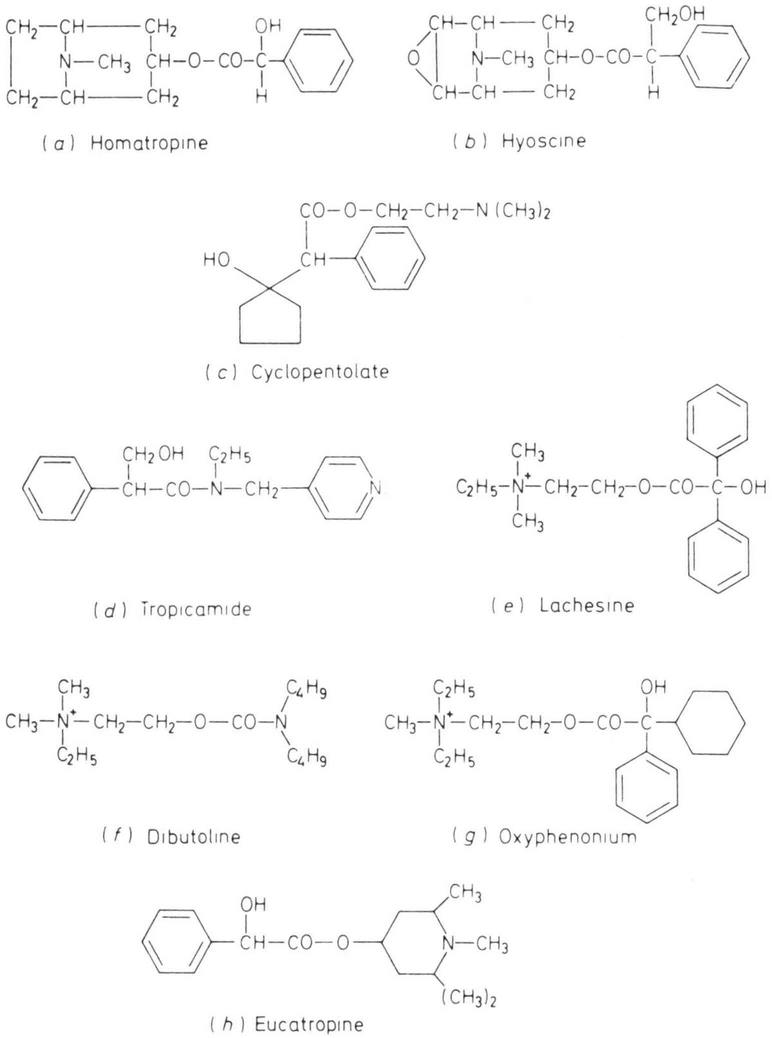

(*a*) Homatropine

(*b*) Hyoscine

(*c*) Cyclopentolate

(*d*) Tropicamide

(*e*) Lachesine

(*f*) Dibutoline

(*g*) Oxyphenonium

(*h*) Eucatropine

Figure 3.2. The structure of some muscarinic antagonists

dropped in the conjunctival sac and applied to the skin, zinc ointment may also be applied to the skin. An antihistamine, which antagonizes most of the actions of histamine, may be prescribed; however, as histamine alone is not responsible for all the symptoms and signs of the allergic response, some of these may persist despite the antihistamine.

Preparations

The alkaloid atropine is almost insoluble in water, preparations therefore contain a salt of atropine, generally the sulphate. There are two official preparations of atropine sulphate.

Eye drops of atropine sulphate, BP Add., 1977

Atropine sulphate	up to 2%
Phenylmercuric nitrate or acetate	0.002%
or	
Benzalkonium chloride	0.01%
Purified water	to 100%

The maximum recommended concentration is 2%, but an optician may prefer to use lower concentrations, for example, 1%. The drops are usually supplied in 10 ml quantities. Single-dose disposable containers of 1% atropine sulphate solution are also available.

Eye ointment of atropine sulphate, BP, 1973

Atropine sulphate	1.0%
Eye ointment basis, BP	to 100%

The ointment, which is the preferred form, is usually supplied in tubes containing 3 g.

The maximum recommended oral dose of atropine sulphate is 2 mg. A 10 ml of 1% eye drops contains 100 mg of atropine sulphate. A 3 g tube of 1% eye ointment contains 30 mg.

Practical aspects

The preparations containing atropine are usually restricted to use in young children. The optician's first task is to decide whether or not a cycloplegic is necessary and therefore carry out a pre-cycloplegic examination. Children selected for refraction under atropine will be those with a high amplitude of accommodation who are not reliable in subjective tests. Atropine may also be used in young children with convergent squint to reveal both the role of accommodation in the squint and the full associated refractive error (usually hyperopia). There is some disagreement concerning the use of atropine in intermittent squint because of the possible danger that the prolonged paralysis of the accommodation may result in its conversion into a constant squint. Other indications for atropine use include high esophoria, where there is a need for the fullest possible correction, and pseudomyopia, in order to reveal the true refractive condition.

Atropine may also be used by the ophthalmologist in certain traumatic conditions and to prevent the development of synechiae in inflammatory states.

There are a number of disadvantages associated with the use of atropine preparations which are listed below:

(1) The full cycloplegic effect of atropine is reached only slowly.
(2) Full dilatation of the pupil may make spherical aberration more noticeable in retinoscopy.
(3) The complete cycloplegia necessitates an allowance for tonus.
(4) Cycloplegia is long lasting.
(5) Mydriasis and the loss of the light reflex may result in discomfort.
(6) Atropine is a strong poison.
(7) Atropine can induce allergic responses.

Atropine may precipitate rises in intraocular pressure in susceptible people, usually the middle-aged or elderly. In this group, however, there is no need for such a potent cycloplegic for refraction.

Homatropine

This is a synthetic compound prepared from mandelic acid and atropine, in which the tropic acid of atropine is replaced by mandelic acid. Homatropine competes with acetylcholine for the receptor sites on smooth muscle. However, homatropine is less potent than atropine and has a shorter duration of action.

Actions in the eye

Sphincter pupillae

Using drops containing 2% homatropine hydrobromide mydriasis commences in 10–15 minutes and a maximum effect may be reached in less than 30 minutes, although the pupil may not be as fully dilated as with atropine. During the maximum effect the response to light is abolished. With the 2% solution mydriasis may remain maximal for a day and two days may be needed for full recovery. If lower strengths are used, the pupil may return to normal within 24 hours.

Ciliary muscle

After the instillation of 2% drops the amplitude begins to fall in about 15 minutes and proceeds rapidly to a maximum effect in

45–90 minutes. Cycloplegia is not usually as complete as with atropine. Recovery time depends on the dosage employed but even with the higher concentrations many patients can resume close work in 5–6 hours.

Intraocular pressure

Because homatropine has the same mechanism of action as atropine it will produce the same effects on intraocular circulation, the iris and canal of Schlemm favouring an increase in intraocular pressure. Therefore, homatropine will suffer the same limitations in use as atropine, but as homatropine has a shorter duration of action, the risk is reduced.

Other actions of homatropine

Homatropine inhibits the action of acetylcholine after it is released from postganglionic cholinergic nerves. Thus, it produces the same pharmacological effects as atropine. Unlike atropine, it is rarely used systemically but is reserved for ophthalmic use.

Preparations

Because the alkaloid is almost insoluble in water the hydrobromide salt is used in preparations. The official preparation is:

Eye drops of homatropine, BP Add., 1977

Homatropine hydrobromide	up to 2%
Benzalkonium chloride	0.01%
or	
Chlorhexidine acetate	0.01%
Purified water	to 100%

The usual quantity supplied is 10 ml. Single-dose disposable units are also available and contain a 2% solution of homatropine hydrobromide. The maximum recommended oral dose of homatropine hydrobromide is 2 mg. A 10 ml bottle of 2% eye drops contains 200 mg.

Practical aspects

Homatropine is weaker than atropine and cannot produce a complete cycloplegia. It is therefore useful in young adults where complete cycloplegia is not necessary and a shorter recovery time is advantageous. Homatropine finds most use in those cases in which a quietening of accommodation allows the true refractive state to be determined, for example, in young people who have restless and spasmodic accommodation, in those patients where the optician suspects a latent error and in pseudomyopia where it is necessary to break down the spasm to reveal the true error.

The disadvantages of homatropine are essentially those of atropine, with the exception that there is a short duration of action and it is not necessary to make an allowance for tonus as cycloplegia is less complete. Caution must be exercised if homatropine is used in elderly patients.

Cyclopentolate

Many atropine-like substances have been synthesized in an attempt to produce compounds active in peptic ulcer. Some were tested in the hope of finding a more ideal cycloplegic, that is, a potent drug with a rapid onset and a short duration of action lacking the toxic potential of atropine. Many were inefficient, painful on instillation or too efficient, having a more prolonged action than atropine. All these were discarded.

Among a few compounds having a short-lived, intense action was cyclopentolate, which has sufficient structural similarity to atropine to make it an effective acetylcholine antagonist. However, the difference between the two structures is such that cross-sensitization between the two drugs does not occur. Cyclopentolate can produce a hypersensitivity reaction of itself but the incidence of this is less than with atropine.

Mydriasis usually commences in a few minutes and the pupil is widely dilated in about 30 minutes.

The onset of cycloplegia occurs at the same time. The fall in amplitude proceeds rapidly, and even in children it is reduced to 1D or 2D in about 20 minutes. The average time to maximum effect is 30 minutes although it may be as much as 60 minutes in some patients. The average duration of the maximum effect is 45 minutes. Recovery from mydriasis and cycloplegia is quite rapid, the normal amplitude may be regained in 24 hours, recovery of original pupil size taking a little longer.

Cyclopentolate is used as the water-soluble hydrochloride salt. The official preparation is:

Eye drops of cyclopentolate, BPC, 1973

Cyclopentolate HCl	up to 1%
Benzalkonium Cl	0.01%
Purified water	to 100%

The drops may also contain boric acid and KCl to buffer the solution to a pH of approximately 5 (cyclopentolate being less stable at room temperature when the pH exceeds 6).

The drops should be stored in a cool place. Single-dose disposable packs are also available containing cyclopentolate HCl 0.5 and 1.0%. As cyclopentolate is not used systemically there is no maximum recommended dose quoted.

Practical aspects

Cyclopentolate may be useful in many of the situations previously discussed under atropine and homatropine. For comparable degrees of cycloplegia it has a shorter recovery period than homatropine and it is not necessary to make an allowance for tonus as the cycloplegia is not complete. It has been suggested, however, that the instillation of 3 drops of 1% cyclopentolate at 10 minutes intervals will produce cycloplegia comparable to that produced by the use of atropine for 3 days. A short period of maximum cycloplegia may be regarded by some as a disadvantage because it reduces the time available for the examination. In cases of atropine allergy the optician could choose either homatropine or cyclopentolate as an alternative cycloplegic. Cyclopentolate may be preferred for two reasons. First, the residual accommodation after cyclopentolate is usually less than that found after homatropine. Second, homatropine is a close chemical analogue of atropine, whereas cyclopentolate has a markedly different structure. The patient allergic to atropine thus is likely to also be allergic to homatropine.

There have been no reports of local toxic effects associated with cyclopentolate, however there are now a number of reports of systemic effects after topical application. These effects are due to an action on the central nervous system causing hallucinations (mainly visual), ataxia, incoherent speech and disorientation. Recovery takes place in some hours with no apparent lasting effects.

As with all cycloplegics there is some danger of inducing raised intraocular pressure in susceptible patients.

Hyoscine

Hyoscine is an alkaloid occurring naturally in certain solanaceous plants. It is used as the hydrobromide salt.

Actions

The actions of hyoscine are similar to those of atropine at the periphery, it blocks the access of acetylcholine to the muscarinic receptors. Centrally it is a depressant; there is no initial stimulation as seen with atropine.

Actions in the eye

The effects of hyoscine are similar to those of atropine except that it has a much shorter duration of action. Using a 0.25% solution the maximum cycloplegic effect occurs within 40 minutes. The amplitude of accommodation has returned to about 4D in 3–4 hours but full restoration may take more than 24 hours. The effect of hyoscine on the sphincter pupillae slightly preceeds that on the ciliary muscle and lasts slightly longer.

Preparations

Eye drops of hyoscine, BPC, 1973

Hyoscine HBr	0.25%
Benzalkonium Cl	0.01%
or	
Chlorhexidine acetate	0.01%
Purified water	to 100%

Single-dose disposable packs are also available and contain 0.2% hyoscine hydrobromide solution.

Eye ointment of hyoscine, BPC, 1973

Hyoscine HBr	0.25%
Eye ointment base	to 100%

The maximum recommended oral dose of hyoscine hydrobromide is 0.6 mg. A 10 ml bottle of 0.25% drops contains 25 mg. A 3 g tube of 0.25% ointment contains 7.5 mg.

Practical aspects

Although more rapid in onset of action than atropine, some variability in the depth of cycloplegia produced in young children has been claimed. However, it has been reported that 2 drops of 0.05% hyoscine can give satisfactory cycloplegia in 1 hour which persists for 6 hours. The child is able to read within 24 hours. Being more potent than atropine there is an increased risk of toxic effects, which casts doubt on the value of hyoscine as an atropine substitute.

Synthetic analogues of atropine

The following four drugs are synthetic analogues of atropine. They are all muscarinic blocking agents and therefore show similar effects to atropine. Dibutoline, oxyphenonium and lachesine do not show the central effects of atropine, their structures presumably making penetration into the central nervous system difficult.

Tropicamide

This compound is of interest because its mydriatic action is reputed to be more marked than its cycloplegic action. Using a 0.5 or 1% solution mydriasis is rapid in onset, reaching a maximum in 15–30 minutes. The pupil is widely dilated. A return to normal occurs within 8–9 hours. Tropicamide has a relatively weak cycloplegic action. The maximum effect is reached after about 25 minutes and is transient. Full amplitude of accommodation has returned at 6 hours. It may be necessary to use repeated instillations to obtain adequate depth and duration of action for refraction. For this reason tropicamide finds more use as a mydriatic than as a cycloplegic.

Lachesine

After instillation of 2 drops of a 1% solution the maximum mydriatic effect is reached in about 1 hour and lasts 5–6 hours. It is a less potent cycloplegic than atropine. The official preparation is:

Eye drops of lachesine, BPC, 1973

Lachesine chloride	up to 1%
Phenyl mercuric nitrate/acetate	0.002%
Purified water	to 100%

This drug may be used in patients with atropine allergy.

Dibutoline

Dibutoline acts as a cycloplegic and mydriatic by the same mechanism as atropine, that is, competitive inhibition of acetylcholine at muscarinic receptors in the sphincter pupillae and the ciliary muscle. Mydriasis and cycloplegia both have a rapid onset. Using a 5% solution, the effects are maximal within 60 minutes and recovery takes 6–12 hours. It is interesting that mydriasis and cycloplegia apparently follow the same time course.

It can therefore be assumed that when mydriasis is complete, cycloplegia is also complete (compare this with atropine and most other cycloplegics). Because of its chemical structure, the compound has the properties of a wetting agent. The resultant effect is an increase in the rate of penetration, thus providing an explanation for the coincidence of the cycloplegic and mydriatic effects.

Oxyphenonium

This is a potent, long-acting compound prepared as the methyl bromide which is water soluble. A 1% solution has similar mydriatic and cycloplegic actions to 1% atropine. It has been particularly recommended where complete cycloplegia is required. Cross-sensitization is unlikely to occur because of the difference in structure.

4

Mydriatics

Mydriatics are used by the optician to dilate the pupil for a more complete examination of the fundus, the vitreous and the periphery of the lens, preferably without accompanying cycloplegia. They are more frequently required in elderly patients because the pupil size decreases with age and because lenticular opacities are more common.

Mydriatics may be useful in young patients when congenital or developmental cataracts prevent a clear view of the fundus or when it is necessary to dilate the pupil to determine the extent and nature of a cataract.

A dilated pupil is essential to examine a fundus thoroughly, and a dilated pupil with abolition of the light reflex is necessary for indirect ophthalmoscopy and slit lamp fundoscopy.

There are two mechanisms by which drugs may produce mydriasis.

(1) Blockade of normal sphincter tone by inhibiting the action of acetylcholine liberated from postganglionic parasympathetic nerves (the muscarinic blocking drugs). Drugs of this type (that is, atropine-like) usually also affect the ciliary muscle which receives the same type of innervation. In order to obtain a selective mydriasis, therefore, it is necessary to choose drugs of low potency relative to atropine or to use lower concentrations of the cycloplegic drugs.

(2) Stimulation of the dilator pupillae will cause a selective mydriasis. Drugs which act in this way mimic the action of noradrenaline liberated from postganglionic sympathetic nerves and are therefore known as sympathomimetic drugs. Sympathomimetic drugs have little or no effect on the ciliary muscle and therefore may be used without the attendant complication of cycloplegia. In the eye these drugs will act on the α-receptors of smooth muscle (that is, peripheral blood vessels and the dilator pupillae).

Precautions with mydriatics

Mydriasis, with or without accompanying cycloplegia, is more frequently required in elderly patients. Predisposition to glaucoma, and therefore the possibility of a rise in intraocular pressure due to blockage of the angle by the iris, is considerably increased in this group of patients. Thus, if mydriasis is required, it is desirable to carry out prior investigations in an attempt to determine whether or not the patient is predisposed. Such investigations might include taking a careful history of the patient, examining the anterior chamber depth and the angle, plotting the fields of vision and measuring the intraocular pressure.

If, after the use of a mydriatic, there is some anxiety about a patient then a miotic may be instilled (*see* Chapter 5).

The muscarinic blocking drugs

Because of their action at the sphincter, which is a more powerful muscle than the dilator, their mydriatic effect will be relatively difficult to overcome and the pupillary response to light will be abolished.

Tropicamide

As mentioned in the section dealing with cycloplegics there is a considerable dissociation between the mydriatic and cycloplegic effects of tropicamide. Use is made of this fact to obtain a good dilatation with only minimal and short-lived effects on accommodation. Using a 0.5% solution maximum mydriasis is achieved in 30–60 minutes and pupil size will return to normal in 8–9 hours. Tropicamide (0.5 and 1%) is available in both single and multidose preparations.

Cyclopentolate

Most of the relevant information regarding cyclopentolate can be found in the section on cycloplegic drugs. A concentration of 0.1% has been recommended for use as a mydriatic when cycloplegia is undesirable. Unfortunately this lower strength is no longer available in the single-dose disposable pack.

Homatropine

Most of the relevant information regarding homatropine can be found in the section on cycloplegic drugs. As a mydriatic it is normally used in solutions containing 0.25–0.5% of the hydrobromide salt. With these concentrations mydriasis commences in 10–20 minutes and the pupil is widely dilated in 30 minutes. At this time the pupil is inactive to the stimulus of light. Accommodation is not markedly affected and recovery from the mydriasis commences in 5–10 hours.

Atropine methylbromide and methonitrate

These quaternary derivatives of atropine are fully ionized at the pH of tear fluid and therefore are relatively poorly absorbed. A 1% solution is required for mydriasis which is maximal between 30 minutes and 1 hour and recovery takes up to 24 hours. For adequate cycloplegia, 5% solutions are required and more than one instillation may be necessary. Even then, the cycloplegia is not as complete as with atropine.

The sympathomimetic drugs

The dilator pupillae is less powerful than the sphincter muscle and therefore the mydriasis is not usually as complete as with the muscarinic blocking drugs. The pupillary reflex to light is slowed and incomplete. The effects of these drugs may be reliably reversed with miotics (*see* Chapter 5).

Adrenaline

Adrenaline is the natural hormone of the adrenal medulla and chemically is closely related to the sympathetic neurotransmitter noradrenaline. The chemical structure of the sympathomimetic drugs is shown in *Figure 4.1*.

Adrenaline does not dilate the pupil when dropped into the conjunctival sac in the concentration of 0.1%, which is the concentration most likely to be used by the ophthalmic optician as a decongestant (vasoconstrictor). If a mydriasis is observed with this concentration then a case of sympathetic hyperirritability is indicated, for example, hyperthyroidism. A pupillary dilatation may occur if the cornea is damaged. A 1 or 2% buffered solution of adrenaline may be used in the treatment of open-angle

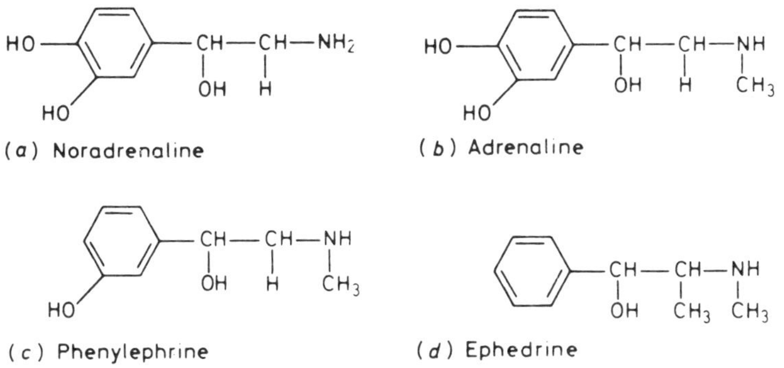

(a) Noradrenaline

(b) Adrenaline

(c) Phenylephrine

(d) Ephedrine

Figure 4.1. The structure of some sympathomimetics

glaucoma. Mydriasis may accompany the use of these higher concentrations.

Phenylephrine

Phenylephrine is closely related in structure to adrenaline (*Figure 4.1*) and is a potent stimulator of the α-adrenoceptor with little effect on the β-receptor. Unlike ephedrine there is no indirect component of the action of phenylephrine.

It is used as the hydrochloride in 10% solution. A 3% solution is isotonic with plasma. The official preparation is:

Phenylephrine eye drops, BPC, 1973

Phenylephrine hydrochloride	10%
Sodium citrate	0.3%
Sodium metabisulphite	0.5%
Benzalkonium chloride	0.01%
Purified water	to 100%

Single-dose disposable units of a 10% solution are also available.

Action in the eye

When instilled as a 10% solution phenylephrine causes a mydriasis which is maximal in 30–60 minutes. Even a concentration of 0.125% used as a vasoconstrictor may cause mydriasis, particularly if the cornea is damaged. The effect is of relatively short duration and may be less marked in patients with heavily pigmented irides. There is little effect on accommodation. A fall in intraocular pressure may be observed, although in elderly patients there is the danger of occluding narrow angles and initiating a rise.

Phenylephrine and other sympathomimetic mydriatics have been observed to release pigment granules from the iris neuroepithelium. It has been suggested that the pigment is released from degenerating cells of the neuroepithelium, which rupture during sympathomimetic-induced contraction. This is most prevalent in elderly people and angiosclerosis may constitute a predisposition to this action.

Ephedrine

The pharmacological effects of ephedrine are similar to those of adrenaline, but are more prolonged. It is used as the hydrochloride salt which is water soluble. A 3.2% solution is isotonic with plasma. Unfortunately, the single dose ophthalmic preparation is no longer available.

Mode of action

Because of the structural similarity to adrenaline it was at first suggested that ephedrine acted directly on the same receptors. However, as its effects were reduced by denervation, it was suggested that ephedrine acted by inhibiting the breakdown of noradrenaline released from sympathetic nerves. At this time, the enzyme monoamine oxidase (MAO) was thought to be the agent responsible for the deactivation of noradrenaline. It is now known that neuronally released noradrenaline is not inactivated by MAO but by the uptake into the nerve and the subsequent storage in an inert complex. Therefore an action on MAO does not explain the sympathomimetic effects of ephedrine. It has been demonstrated that ephedrine has a two-fold action at sympathetically innervated tissue. First, it acts directly on the receptor (α or β), an action constituting about 20% of the final response. The remaining 80% of the response is an indirect effect due to an uptake of ephedrine into sympathetic nerves and release of the stored noradrenaline from these nerves. The released noradrenaline then acts on the closely associated receptor.

Action in the eye A 5% solution of ephedrine hydrochloride begins to dilate the pupil in 5–10 minutes after instillation. The mydriasis is usually sufficient for fundus examination within 20–25 minutes. The maximum mydriatic effect may take longer than 30 minutes.

Some patients may prove resistant to the actions of ephedrine. In older patients with arteriosclerosis the radial blood vessels of the iris have become hardened and are resistant to the pull of the relatively weak dilator muscle. In these cases one of the muscarinic blocking agents may be required or a mixture of the two types of mydriatics may be preferred. Some resistance to ephedrine may also be seen in patients with heavily pigmented irides.

Although there is no good evidence for the involvement of the sympathetic system in the control of accommodation, it has been reported that ephedrine causes some reduction in the amplitude of accommodation within the range 0.5–2D.

A 5% solution of ephedrine hydrochloride may be used in anisocoria (unequal pupil size) to determine whether the situation is pathological or physiological.

Young patients show no significant change in intraocular pressure although there may be some tendency to a slight reduction. A similar effect is seen in older patients with normal anterior chambers but if the angle is narrow and the anterior chamber shallow there is some possibility of a rise.

Cocaine

Cocaine is dealt with in detail in the section on local anaesthetics. However, in addition to its local anaesthetic actions, cocaine is a mydriatic and vasoconstrictor drug. This is a property peculiar to cocaine and not shared by the other local anaesthetics. Because of its addictive liability, cocaine is controlled and not available to the ophthalmic optician.

Mode of action

Cocaine is an indirectly acting sympathomimetic agent. It inhibits the active uptake of noradrenaline into postganglionic sympathetic nerves. Thus, noradrenaline released from the nerves remains in the synaptic region and reaches concentrations sufficient to activate the α-receptors of the smooth muscle of the iris or peripheral blood vessels. There is no direct component of action.

Action in the eye After instillation of a 2% solution of cocaine mydriasis commences in 5–20 minutes and recovery may take longer than 6 hours. There is little effect on accommodation. The light reflex is still present but reduced. Because of its systemic toxicity and its addictive liability cocaine is not used as a mydriatic agent by the ophthalmic optician, although the ophthalmologist may find cocaine of value, for example, in combination with homatropine.

Amphetamine and hydroxyamphetamine

These drugs are indirectly acting sympathomimetics. Their mode of action is similar to ephedrine. Solutions of 1% give mydriasis with no effect on accommodation. However, they have no advantages over ephedrine and phenylephrine. The dangers of amphetamine abuse make it desirable to limit its use as much as possible. Therefore it is not recommended or available for use by the ophthalmic optician.

5

Miotics

Stimulation of the parasympathetic nerve supply to the iris causes release of acetylcholine and contraction of the sphincter pupillae muscle (miosis). Acetylcholine itself is ineffective as a miotic when instilled into the eye, as it is rapidly broken down by the enzyme acetylcholinesterase. Thus, other drugs must be used to produce miosis.

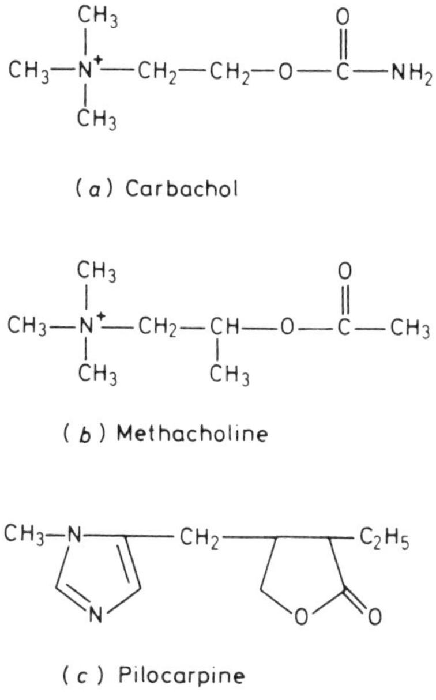

$$CH_3-N^+(CH_3)_2-CH_2-CH_2-O-C(=O)-NH_2$$

(*a*) Carbachol

$$CH_3-N^+(CH_3)_2-CH_2-CH(CH_3)-O-C(=O)-CH_3$$

(*b*) Methacholine

(*c*) Pilocarpine

Figure 5.1. The structure of some direct acting miotics

There are three types of miotics in use: direct-acting miotics which act like acetylcholine but are stable and resistant to cholinesterases (*Figure 5.1*), anticholinesterases (*Figure 5.2*) and α-receptor antagonist, thymoxamine (*see* p.51).

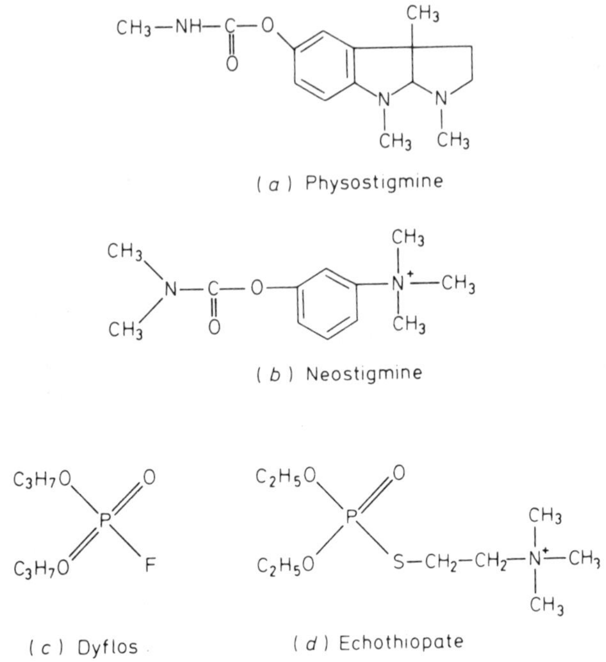

(a) Physostigmine

(b) Neostigmine

(c) Dyflos

(d) Echothiopate

Figure 5.2. The structure of some anticholinesterase miotics

Direct-acting miotics

These are of two types: the cholinomimetic alkaloids and the choline esters.

Pilocarpine

Pilocarpine is a cholinomimetic alkaloid obtained from the leaves of a South American shrub, *Pilocarpus jaborandi.* The base is a colourless liquid of syrupy consistency, soluble in water. The official salt is the hydrochloride. (Other cholinomimetic alkaloids are muscarine and arecoline but these are not available for ophthalmic use.)

Pilocarpine has the muscarinic actions of acetylcholine because it is able to occupy the muscarinic receptors present at the junction of postganglionic parasympathetic nerves and their effector organs. It is resistant to the hydrolytic action of the cholinesterase enzymes and thus has a useful duration of action. It has relatively little activity at the nicotinic receptors of the ganglia and skeletal muscle. If pilocarpine enters the systemic circulation by absorption from the conjunctival or nasal mucous membranes then the effects seen will be those of stimulating muscarinic receptors and will include an increase in the secretions of the salivary, lacrimal, gastric and pancreatic glands. There will also be increased sweating, an increase in the tone and motility of the gastrointestinal tract, bronchoconstriction and slowing of the heart. In the unlikely event of overdosage, all the above effects will occur together with nausea, vomiting, hypotension, tremor and if not treated, death due to respiratory collapse.

Actions in the eye

Sphincter pupillae With the 1% solution miosis commences in a few minutes and remains marked for about 6 hours. Once the drug has diffused away the pupil may become larger than normal due to a diminished response of the sphincter to reflex stimulation.

Ciliary muscle The spasm of accommodation lasts about 1 hour and once recovery begins, attempts at close work do not cause the spasm to return. This is an advantage over physostigmine (*see* below).

After instillation pilocarpine causes dilatation of the conjunctival vessels which usually disappears in 30 minutes. Toxic effects after instillations of 1 or 2 drops are uncommon but an allergy may occur if the use is prolonged, as in glaucoma.

Preparations

Eye drops of pilocarpine, BP Add., 1977

Pilocarpine hydrochloride	up to 4%
Benzalkonium chloride	0.01%
Purified water	to 100%

Pilocarpine is also available as a 1, 2 and 4% solution of pilocarpine nitrate in single-dose disposable containers.

The maximum recommended oral dose is 12 mg. A 10 ml bottle of 4% eye drops contains 400 mg pilocarpine hydrochloride.

The choline esters are a group of compounds derived from choline and having similar structures to acetylcholine. They differ from acetylcholine in being more stable and less susceptible to the cholinesterase enzymes. They act directly at the muscarinic receptor and therefore their actions are qualitatively similar to those of pilocarpine. The choline esters have varying ratios of activity at the muscarinic and nicotinic receptors.

Methacholine

A 10% solution is usually required to produce miosis. Lower concentrations are effective if the corneal epithelium is damaged or if a wetting agent is included in the formulation.

Carbachol

This drug has a high nicotinic component in its action, thus part of its miotic effect may be due to a release of acetylcholine from postganglionic parasympathetic nerves. It will also act directly on the muscarinic receptor. Carbachol is virtually resistant to the cholinesterase enzymes and therefore has a sufficiently long duration of action to be useful. Because carbachol has low lipid solubility at any pH it has poor penetrating powers. The presence of benzalkonium chloride in the formulation improves absorption and enhances the drug action.

In concentrations of 0.5–2% carbachol produces a powerful miosis which lasts 3–8 hours, together with spasm of accommodation which is more severe than with pilocarpine and may cause headache. Instillation of carbachol may produce local vasodilatation. Allergic reactions are uncommon.

There is an official preparation:

Eye drops of carbachol, BPC, 1973

Carbachol	up to 3%
Benzalkonium chloride	0.01%
Purified water	to 100%

The maximum recommended oral dose is 4 mg. A 10 ml bottle of 3% eye drops contains 300 mg.

Anticholinesterases

Anticholinesterases form complexes with cholinesterase enzymes and thereby prevent hydrolysis of acetylcholine (*Figure 5.3*). They are divided into reversible and irreversible inhibitors.

Reversible anticholinesterases have some similarity in structure to acetylcholine. They combine with the ionic and esteratic sites of the cholinesterase enzyme but are not readily susceptible to hydrolysis.

Physostigmine

This is a naturally occurring compound extracted from the Calabar bean (dried ripe seeds of *Physostigma venenosum*). The pure

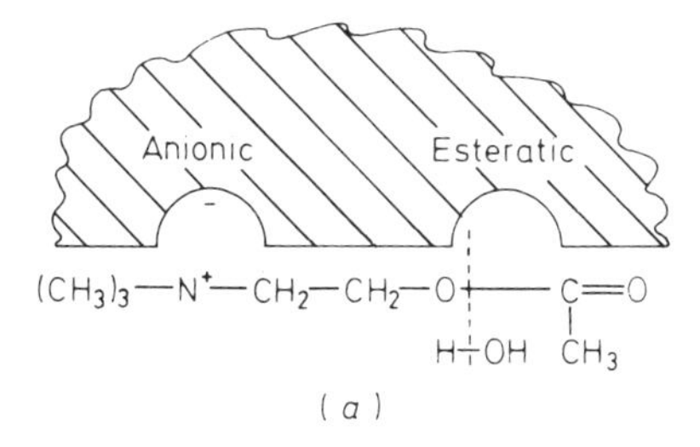

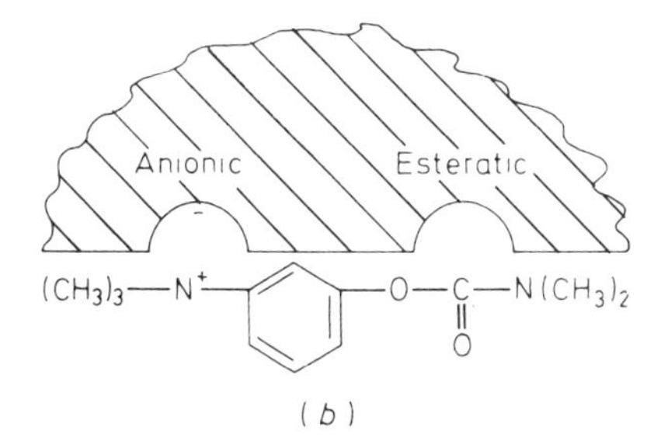

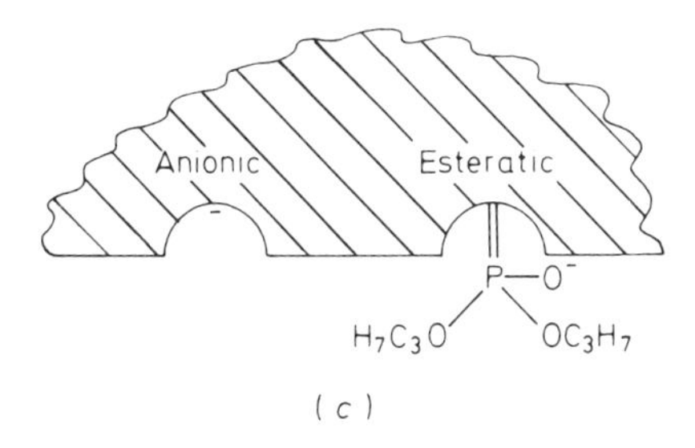

Figure 5.3. Schematic diagram showing the interaction of acetylcholine, neostigmine and dyflos with acetylcholinesterase. The acetylcholinesterase enzyme has anionic and esteratic sites. (*a*) Acetylcholine binds to both sites and is rapidly hydrolysed. (*b*) Neostigmine binds to both sites and is slowly hydrolysed, therefore the enzyme is reversibly blocked. (*c*) Dyflos phosphorylates the esteratic site, the enzyme is irreversibly blocked

alkaloid was isolated independently by two groups of people, one group naming it physostigmine, the other eserine. These two names are both in common use. The systemic effects of physostigmine are equivalent to excessive stimulation of all cholinergic nerves. Thus it will potentiate the actions of acetylcholine at the muscarinic receptor and at the nicotinic receptors of ganglia and skeletal muscle. If sufficient physostigmine is absorbed after topical application to the eye, the symptoms will include salivation, lacrimation, bronchoconstriction, increased tone and motility in the gastrointestinal tract, increased secretion from mucous glands, possibly nausea and vomiting, slowing of the heart and fasciculation in skeletal muscle followed by muscle weakness. With very high doses, effects on the central nervous system occur which include confusion, ataxia, loss of reflexes, convulsions and central respiratory paralysis.

Actions in the eye

Sphincter pupillae Using 1% solution of physostigmine sulphate pupillary constriction begins in less than 10 minutes and proceeds rapidly so that the pupil diameter is reduced to less than 2 mm. Marked constriction may persist for more than 12 hours and complete recovery may take a number of days.

Ciliary muscle The effects of physostigmine on the ciliary muscle begin after the onset of miosis. The ciliary muscle contracts to produce an artificial myopia. The spasm lasts for 2–3 hours. If close work is attempted after this time the spasm may recur. This is because there is sufficient cholinesterase inhibition remaining to cause potentiation of the acetylcholine liberated during the attempt to focus. This feature may remain for several hours.

After instillation physostigmine causes dilatation of the conjunctival vessels accompanied by stinging. The intense spasm of accommodation may result in severe discomfort and the patient often complains of pain over the brow. There may be a twitching of the lids and spasm of the extra-orbital muscles due to potentiation of acetylcholine at the nicotinic receptors. An allergic reaction may develop after prolonged use.

There is an official preparation:

Eye drops of physostigmine, BP Add., 1977

Physostigmine sulphate	up to 1%
Sodium metabisulphite	0.2%
Benzalkonium chloride	0.01%
Purified water	to 100%

There is also a combined preparation of physostigmine and pilocarpine:

Eye drops of physostigmine and pilocarpine, BPC, 1973

Physostigmine sulphate	up to 0.5%
Pilocarpine hydrochloride	up to 4.0%
Sodium metabisulphite	0.2%
Benzalkonium chloride	0.01%
Purified water	to 100%

Physostigmine is relatively unstable and in the absence of the antoxidant sodium metabisulphite rapidly turns pink to deep red, due to the formation of rubreserine. This product of oxidation is also a miotic compound but is irritant. Sodium metabisulphite delays the appearance of the red oxidation product partly by a decolorizing action. There is thus a limited storage period for physostigmine which should be observed even if the solution has not discoloured.

Neostigmine

This has a similar mode of action to physostigmine. It is highly ionized in aqueous solution and therefore is relatively poorly absorbed after topical application and the response may vary from patient to patient. No official preparation is available, proprietary preparations of 3 to 5% have been used in the treatment of glaucoma.

Irreversible or *long-acting anticholinesterases* have an extremely long duration of action. They are usually organophosphorus compounds which act at the esteratic site of the cholinesterase enzyme (*Figure 5.3*). Cholinesterase activity can only be restored by synthesis of new enzyme material, which explains their prolonged duration of action. These compounds are potent and extremely toxic and should be used with caution under medical supervision. They include dyflos and echothiopate. The organophosphorus compounds are more efficient inhibitors of cholinesterase than acetylcholinesterase, unlike the reversible anticholinesterases which inhibit both types of enzymes equally.

Dyflos

Dyflos is a colourless oily liquid which is volatile and unstable in water. It may be used in glaucoma as a 0.1% solution in arachis oil. It causes pronounced miosis and ciliary spasm.

Ecothiopate

Ecothiopate is a compound with a similar type of action to dyflos but it is more stable in aqueous solution, with a half-life of days rather than hours. It may be prepared in aqueous solution as the iodide salt. Benzalkonium chloride, which is incompatible with iodides, should not be used as a preservative. There is no official preparation but it is marketed in proprietary preparations containing up to 0.25%.

As well as the dangers of systemic effects due to absorption after instillations, these substances have two other serious disadvantages if their use is prolonged. There is the danger of delayed neurotoxicity, nerves become demyelinated and there is a spastic paralysis followed by a flaccid paralysis in skeletal muscle. The second disadvantage is the possibility of long-term use causing lens opacities.

A compound with a long duration of action which is not of the organophosphorus type is demecarium. This substance is synthesized from two molecules of neostigmine connected together by a carbon chain. It is water-soluble and stable. This drug will produce a miosis which is maximum in 2–4 hours and is sustained for 5–7 days after instillation of a 0.5% solution. After systemic absorption, side-effects are those that would be expected of cholinesterase inhibition.

Use of miotics to reverse mydriasis

When a mydriatic drug has been used for fundus examination it may be considered desirable to reverse the mydriasis. This is important if it is suspected that the patient has a predisposition to glaucoma, when a prolonged mydriasis could precipitate a rise in intraocular pressure. The miotics generally used are pilocarpine 1% and physostigmine 0.5% or 0.25%. Of these two, pilocarpine is probably the agent of choice as there is likely to be less pain.

There are a number of drawbacks to the use of miotics and the indiscriminate reversal of mydriasis is not recommended. Thus, in younger patients, the miotic effect and spasm of accommodation may be prolonged and cause more discomfort than would have been experienced with the mydriatic alone. This is most likely to occur if the sympathomimetic group of mydriatics has been used. In other cases when the long-acting cycloplegic drugs (that is, atropine and high concentrations of homatropine) have been used, there may be no reversal with either pilocarpine or physostigmine. A more potent miotic is required so that if treatment is desired, referral is necessary. If drugs of lower potency are used (cyclopentolate, tropicamide, low concentrations of homatropine) then the result of miotic instillations is variable. In some cases satisfactory reversal may occur. In others the effect of the mydriatic may outlast the effect of the miotic. If there were a predisposition to glaucoma then a second instillation of a miotic would be desirable.

Some of the effects described above are explicable if the differing mechanisms of action of the drugs are understood. The sympathomimetic drugs act on the weak dilator muscle. The miotics act on the stronger sphincter muscle and therefore the mydriasis is relatively easily overcome. It has been suggested that there is pain after the instillation of miotics due to the opposition between the sphincter and dilator muscles. If this is the case, then as both pilocarpine and physostigmine act by muscarinic receptor stimulation (pilocarpine directly, physostigmine indirectly), there is no advantage in the use of one or the other in this respect. The problems encountered with the reversal of sympathomimetic drugs can be avoided by the use of thymoxamine. This drug is a competitive antagonist at α-adrenoceptors and reversal of mydriasis will thus involve action at receptors in only one iris muscle, namely the dilator. In addition, there will be no action at the ciliary muscle. Thymoxamine is available as a 0.5% solution in single-dose disposable packs which should be stored in the refrigerator. There may be some initial transient stinging on instillation of the drops.

The cycloplegic drugs produce mydriasis by a blocking action at the muscarinic receptor on the sphincter muscle. Reversal of this effect relies on a competition between the mydriatic and the miotic agent for these receptors. Drugs with high affinity for the muscarinic receptor, for example, atropine are difficult to reverse because it is not possible to achieve a high enough concentration of the miotic agent at the receptor. In the case of pilocarpine the miotic agent is the drug itself. In the case of physostigmine the miotic agent is the accumulated acetylcholine resulting from cholinesterase inhibition. With cyclopentolate and similar drugs the efficiency of reversal is a complex situation depending on the affinity of the cycloplegic and of the miotic agent for the receptor, the rate of absorption of both types of drugs and their respective time courses of action.

It may occasionally happen that it is desired to use a mydriatic in an elderly patient who has a predisposition to glaucoma. The sympathomimetic drugs used alone may not produce full mydriasis. A combination of ephedrine (5%) and homatropine (0.25%) can be used in this case to give a good mydriasis which is easily reversed because of the low concentration of homatropine. This low concentration is sufficient to cause some block of the sphincter muscle and prevent it opposing the contraction of the dilator muscle. Such a combination may also be of use in patients with heavily pigmented irides if mydriasis without cycloplegia is desired.

6

Local anaesthetics

Local anaesthetics are used to reversibly block pain sensation in relatively restricted areas of the body. When applied in effective concentrations they are able to block nerve conduction in all parts of the nervous system and in all types of nerve fibres (both motor and sensory nerves). They will also block conduction in the heart and other conducting tissues. Sensitivity to touch, pressure, heat and cold is also abolished.

Methods of application

Topical anaesthesia

The local anaesthetic is applied to a surface as a solution, ointment, cream or powder.

Infiltration anaesthesia

The local anaesthetic is injected by the subcutaneous route in order to affect the fine sensory nerve branches.

Nerve block anaesthesia

The local anaesthetic is injected near to a major nerve trunk. Block of conduction of impulses in motor and sensory nerves results. There is thus muscle relaxation as well as loss of sensation.

Spinal anaesthesia

The local anaesthetic is injected into the subarachnoid space producing analgesia and muscle relaxation without the patient losing consciousness.

Properties of an ideal local anaesthetic

(1) Rapid onset of action.
(2) Profound depth of anaesthesia.
(3) Adequate duration for the purpose required.
(4) Reversible action.
(5) No pain at site of administration.
(6) No after pain when the effects wear off.
(7) No hypersensitivity or allergic reactions.
(8) No local or systemic toxicity.
(9) Compatible with any other drug used, including preservatives.
(10) Preferably water-soluble, stable in solution and on autoclaving.

Further requirements for use in the eye

(1) Good penetration.
(2) No irritant or damaging effect on the cornea.
(3) No pupil effect.
(4) No vasodilation.

Uses in ophthalmic optics

(1) Moulding technique in contact lens fitting.
(2) Tonometry.
(3) Removal of foreign bodies.
(4) Gonioscopy.

There are certain dangers associated with the use of local anaesthetics in the eye. If a patient is allowed to leave the consulting room before recovery from the effects of a local anaesthetic then damage may unwittingly be caused as the anaesthetized cornea cannot detect the presence of a foreign body. Repeated instillations of local anaesthetics may result in desquamation of the corneal epithelial cells because normal functioning of the nerves is needed for their survivial. These effects on the cornea are aggravated by the loss of the protective blink reflex and drying of the eye. This may result in pitting and ulceration of the cornea.

Mode of action

Before discussing how local anaesthetics produce their effect, it is necessary to review the current concepts on how a nerve impulse is generated and transmitted.

A nerve is composed of a bundle of nerve fibres, each nerve fibre (*Figure 6.1*) consisting of the following:

(1) The axon, a core of semi-fluid gelatinous material.
(2) A protein-lipid membrane which surrounds the axon.
(3) In some fibres, a myelin sheath which surrounds the membrane. This is interrupted at intervals by the nodes of Ranvier, which are gaps in the myelin sheath.

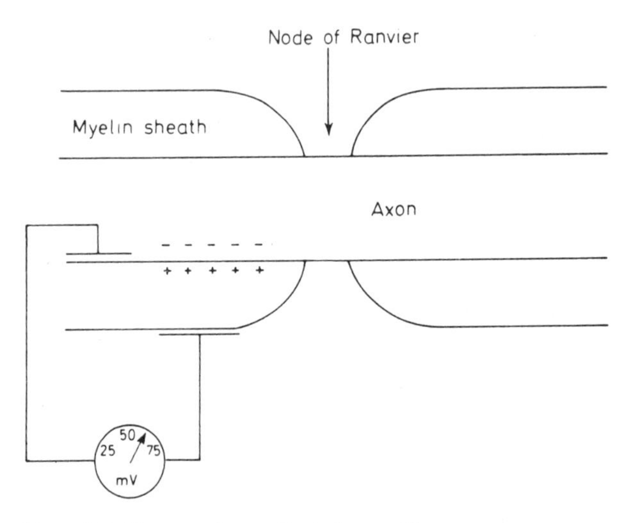

Figure 6.1. Diagram of a myelinated nerve fibre

The inside of the nerve fibre is electrically negative with respect to the outside (*Figure 6.1*). This potential difference is about 50–70 mV and is known as the membrane potential. A potential exists because there is a slight excess of negative ions within the cell and a slight excess of positive ions without (*Figure 6.2*). This situation arises because:

(1) The resting nerve cell membrane exhibits differences in permeability towards various ions, in particular Na^+ and K^+. There is a very low permeability to Na^+ ions.
(2) There is a Na^+/K^+ pump in the membrane which keeps Na^+ outside the nerve and brings K^+ ions into the nerve. The pump requires energy from metabolic processes.
(3) The impermeability of the membrane to certain anions, for example, sulphate and phosphate, maintaining them (A^-) within the cell.

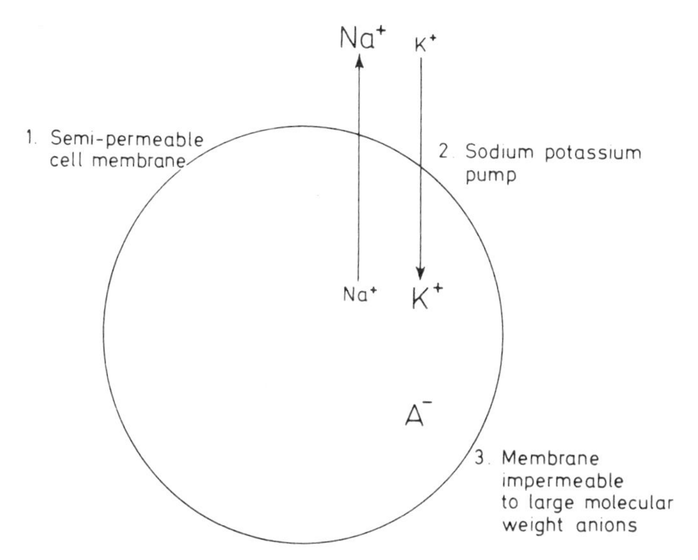

Figure 6.2. The distribution of anions and cations across a nerve cell membrane

If a potential difference were applied between the extracellular fluid and the cytoplasm, it would not be transmitted very far along the fibre but for the fact that the fibre exhibits threshold behaviour, that is, electrical instability. The ionic differences responsible for the membrane potential constitute a store of potential energy and this is made use of in threshold behaviour. If a sub-threshold impulse is applied, there is a small depolarization of the cell (inside of the cell becomes less negative) which rapidly dies away (*Figure 6.3*). If the strength of the applied impulse is increased, the cell is depolarized to the threshold level and a propagated action potential occurs. This travels the length of the nerve fibre without any decrease in magnitude. The threshold is about 20 mV lower than the resting potential. Once the stimulus is sufficient to depolarize the cell to the threshold, increasing it has no effect on the size of the propagated action potential. It can be shown that during the action potential there is a small increase in the Na^+ ion concentration inside the fibre and a small decrease in the K^+ ion concentration. There is an increase in the membrane permeability to Na^+ and this ion enters during depolarization. At the peak of the action potential, about +30–40 mV, the membrane permeability to Na^+ ions decreases and the permeability to K^+ ions increases. These two factors are responsible for repolarization. In this situation K^+ is lost from the nerve. In fact, the nerve

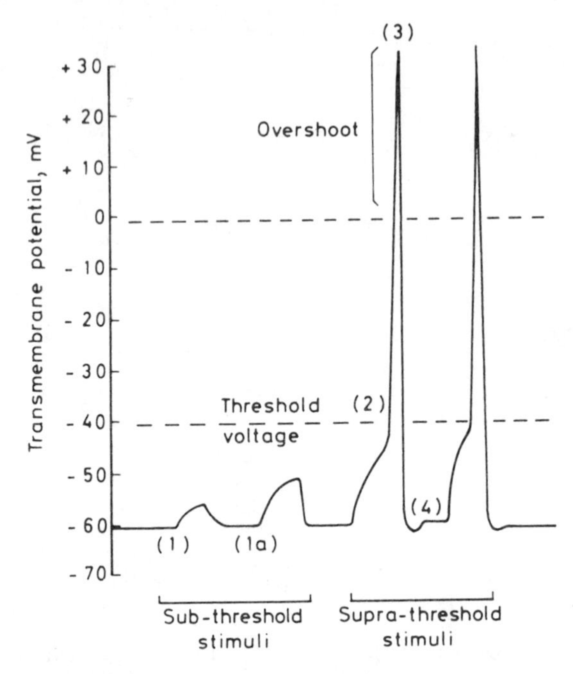

Figure 6.3. Changes in the transmembrane potential of a nerve fibre due to sub-threshold and supra-threshold stimulation. (1) and (1a) Sub-threshold stimuli. (2) At this point the large increase in the permeability to sodium occurs and the sodium ions enter the nerve fibre causing a reversal of polarity. (3) At this point the increase in permeability to potassium occurs and the potassium ions leave the nerve fibre tending to restore the nerve transmembrane potential. (4) At this point the nerve has returned to its resting state, it is relatively impermeable to sodium and potassium ions and the sodium pump is effective

becomes slightly hyperpolarized because the permeability to K^+ ions is still above that found in the resting nerve when the permeability to Na^+ has returned to normal.

Conduction

An action potential is sufficient to supply a supra-threshold stimulus to adjacent regions of the membrane. Local circuit currents are generated, passing from the active to the inactive region within the cell, and from the inactive to the active outside the cell. In this way an action potential is propagated along the nerve fibres. In myelinated nerves the changes in membrane permeability and thus the action potentials occur only at the nodes of Ranvier, due to the insulating effect of the myelin at other sites.

Thus, the depolarizations pass from node to node. There is therefore an increase in conduction velocity. This type of conduction is called saltatory conduction.

An understanding of how local anaesthetics act comes predominantly from studies on easily accessible, large, myelinated fibres described above. Sensory nerve endings, such as those found in the cornea, are non-myelinated. These nerves have not been so well studied. However, from experiments on the myelinated fibres, one can predict how local anaesthetics will affect nerve impulse generation and conduction in fine sensory nerves. In the following discussion it is assumed that the action of local anaesthetics on membranes is the same in both types of nerve.

Local anaesthetics prevent both the generation and conduction of nerve impulses. The site of action is the membrane. Here they prevent the large, transient increase in the permeability to Na^+ ions and also reduce the resting permeability to K^+ ions. The local anaesthetic becomes bound to the outer layers of the membrane, but it is not clear how this binding stabilizes the permeability of the membrane. One suggestion is that local anaesthetics increase the surface pressure of the lipid layer of the membrane, this could close pores through which the ions move. Such an action would result in a general decrease in resting permeability and limit the increase in membrane permeability to Na^+ ions. This suggestion is based on observations that the anaesthetic potency of a series of compounds parallels their ability to increase the surface pressure of artificial monomolecular lipoid films.

Another suggestion involves the competition between the cationic form of the local anaesthetic and calcium ions for receptors involved with the movement of sodium ions. From *in vitro* studies it seems possible that the receptor is located on a phospholipid, that calcium is normally bound to this site and that its release is associated with increased permeability to sodium ions. If this is the case, and the calcium is replaced by local anaesthetic which is not released, then the increased permeability will not take place and conduction in the nerve will be blocked.

Structure and chemistry

There is a great deal of similarity in the structure of local anaesthetics. Most local anaesthetics are tertiary amines (that is, ammonia, NH_3, with all three hydrogen atoms replaced by other groups) but a few are secondary amines. They have the general structure: lipophilic portion; linkage; hydrophilic portion (*Figure 6.4*). The structure of individual local anaesthetics is shown in *Figure 6.5*.

The lipophilic portion gives lipid solubility, the hydrophilic portion water solubility. As discussed in the section on drug absorption, a good balance of these two properties is essential for absorption into the layers of the cornea. The hydrocarbon chain

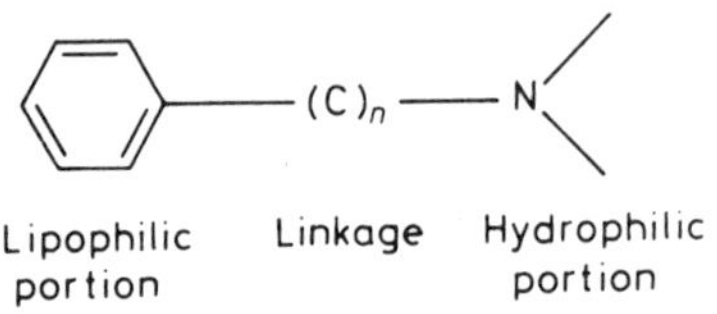

$(C)_n$ — N

Lipophilic portion Linkage Hydrophilic portion

Figure 6.4. The general structure of local anaesthetics

connecting these two portions contains an ester or amide linkage and maintains correctly the fairly critical distance between the lipophilic and hydrophilic portions. A chain length equivalent to 2 or 3 carbon atoms is the best length for this linkage. The type of linkage can affect metabolism and therefore the duration of action.

The ester link

$$-\overset{\overset{\displaystyle O}{\|}}{C} - O - CH_2 -$$

as in procaine, is easily hydrolysed by the enzyme cholinesterase (non-specific) of plasma. Local anaesthetics with unprotected ester linkages therefore have a short duration of action.

The amide link

$$NH - \overset{\overset{\displaystyle O}{\|}}{C} - CH_2 -$$

as in lignocaine, is much more resistant to hydrolysis and therefore local anaesthetics with this link are likely to be longer acting.

Local anaesthetics are weak bases, relatively insoluble in water and rather unstable. They readily form water-soluble salts of strong acids which are stable in solution. Local anaesthetics are therefore generally supplied as the hydrochloride salt. The

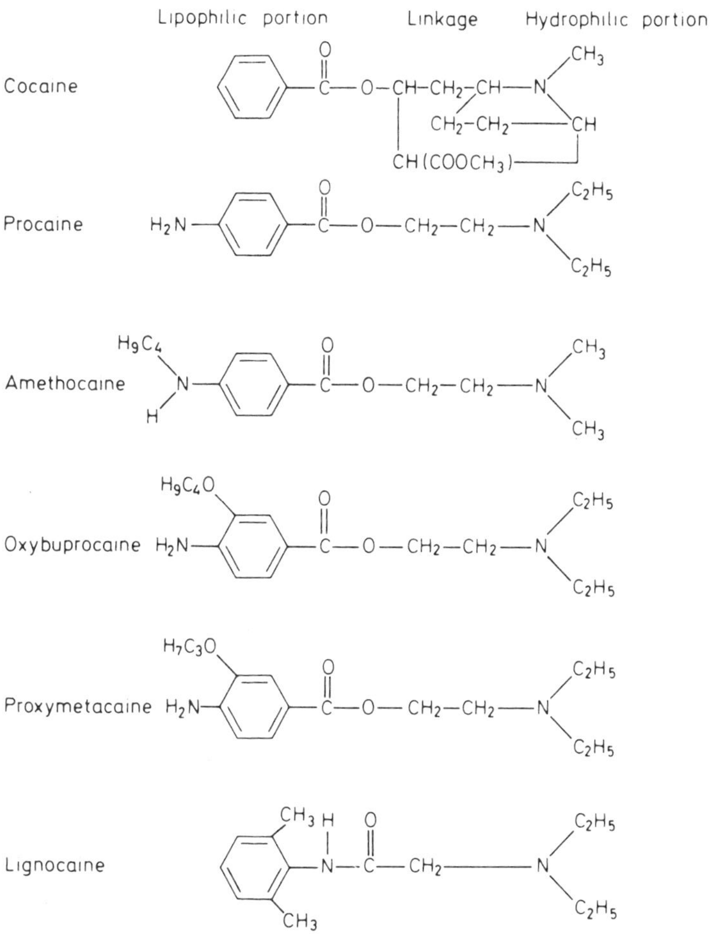

Lipophilic portion Linkage Hydrophilic portion

Cocaine

Procaine

Amethocaine

Oxybuprocaine

Proxymetacaine

Lignocaine

Figure 6.5. The structure of individual local anaesthetics

un-ionized form is important for absorption. However, it is the ionized form, the cation, which is responsible for the local anaesthetic effect. (For a more detailed discussion of this aspect *see* Appendix.)

Onset and duration of anaesthesia

Generally, small nerve fibres are more susceptible to local anaesthetics than are large fibres, and unmyelinated nerve fibres are more susceptible than myelinated fibres, the myelin providing a further barrier to absorption. Susceptibility does not depend on the type of fibre (that is motor or sensory).

For a given nerve fibre the time to the onset of action depends on the concentration of the local anaesthetic, the diffusion time, the nerve diameter and the pH of the solution used (this last affects ionization and therefore absorption, *see* Appendix). Recovery time depends on the bond between the local anaesthetic and the membrane, the fibre size and the rate of removal of the drug. The rate of removal depends, in turn, on factors such as the vascularity of the area and the ease of hydrolysis of the drug. The higher the blood supply, the higher the rate of removal. Some local anaesthetics, for example, procaine, are vasodilators and easily removed from the site, others such as cocaine are vasoconstrictors and are retained at the site.

Absorption, metabolism and excretion

Even after topical application some local anaesthetics will enter the blood. They are removed from the blood by tissue redistribution, metabolic inactivation and excretion. Ester-linked compounds are readily hydrolysed, amide-linked compounds are less easily hydrolysed. Further metabolism takes place in the liver and the metabolites and unchanged drug excreted through the kidneys.

Toxicity

Allergic reactions to the ester-linked compounds are not uncommon. These reactions may take the form of a dermatitis or asthma. Reactions to the amide-linked compounds are rare.

It is unusual to see systemic toxic effects of local anaesthetics, other than cocaine, after local application. The advent of systemic toxicity depends on the elevation of the blood levels and the length of time for which they are elevated. Most conspicuous reactions are convulsions (a central nervous system effect), but the most dangerous are respiratory depression and hypotension. Cocaine is unique in causing marked cortical stimulation. Some people are especially sensitive to cocaine and cardiovascular collapse may occur after small doses.

Cocaine

Cocains is an alkaloid which occurs naturally in the leaves of *Erythroxylon coca.* Its possession is controlled by the law because of its addictive liability. Thus, it is not available to the non-medical optician. However, as it was the first local anaesthetic used it may serve as a standard with which to compare the synthetic local anaesthetics.

It is potent and well absorbed after topical application. It is normally used as 2% drops which give complete anaesthesia in 10–15 minutes. Partial anaesthesia may remain for about an hour. Cocaine may damage the cornea, causing pitting and ulceration. Because it potentiates noradrenaline it causes mydriasis, and vasoconstriction, which may be of use in ophthalmic surgery, providing a bloodless field. Potentiation of adrenaline liberated from the adrenal medulla may be the major factor in cocaine's toxic effect on the heart. Because of the danger of corneal damage, toxicity and addiction, it is not much used.

It may be seen from the above statements that cocaine does not conform to the properties of an ideal local anaesthetic, as listed at the beginning of this chapter, and that alternatives are desirable.

Procaine

Procaine was the first of the synthetic local anaesthetics to come into wide use. However, it is of little value for topical application. It is only poorly absorbed and its action is slow to develop. Surgeons may use it by sub-conjunctival injection as a 5–10% solution. When used in infiltration anaesthesia it is given with a vasoconstrictor such as adrenaline to prolong its action. It is hydrolysed to para-aminobenzoic acid and therefore should be avoided in patients receiving sulphonamide treatment (*see* Chapter 8).

Amethocaine

Amethocaine has been widely used by opticians. It is well absorbed; 0.5–1.0% drops give adequate corneal anaesthesia. The blink reflex is absent after 10–15 seconds. The duration is sufficient for moulding, tonometry or foreign-body removal. Local responses to amethocaine are rare but superficial corneal epithelial lesions may occur after prolonged or repeated use. There is some initial stinging after instillation of the drops. The official formulation is:

Amethocaine BPC, 1973

Amethocaine hydrochloride	up to 1%
Sodium metabisulphite	0.1%
Phenyl mercuric nitrate or acetate	0.002%
Purified water	to 100%

It is also available as 0.5 and 1% solution in single-dose disposable containers.

Oxybuprocaine

Oxybuprocaine is well absorbed. A single instillation of 0.3 or 0.4% gives sufficient anaesthesia for tonometry in less than 60 seconds. With a second drop at 90 seconds anaesthesia is sufficient for contact lens work. A third drop after a further 90 seconds gives anaesthesia which is sufficient for foreign-body removal. Full recovery from three drops occurs in one hour. There is no prior irritation, no action on blood vessels or pupil and no evidence of corneal damage. Some proprietary preparations of oxybuprocaine contain chlorhexidine acetate as a preservative. This preservative is incompatible with fluorescein because a dense precipitate forms on mixing.

Oxybuprocaine is available at 0.4% in single-dose disposable units which contain no preservative.

Proxymetacaine

Proxymetacaine is well absorbed; the onset, depth and duration of anaesthesia are similar to those for oxybuprocaine. Recovery from 1 to 2 drops of a 0.5% solution takes about 15 minutes. No prior irritation occurs and there is no evidence of corneal damage. The drops discolour on exposure to air with an accompanying loss of potency. Drops should be stored in the refrigerator after opening but not allowed to freeze. The preparation should also be protected from light.

Lignocaine

A 4% solution of lignocaine is effective at the cornea, giving a more rapid, intense, extensive and prolonged effect than an equal concentration of procaine. There is no effect on the pupil.

Factors which contribute to the longer duration of local anaesthetic action are the absence of marked vasodilation and a resistance of the compound to hydrolysis. Lignocaine is resistant to hydrolysis because it has an amide rather than an ester link, and because of two methyl substituents on the benzene ring of the lipophilic moiety. These groups prevent the close approach of the drug to the active sites on the cholinesterase enzyme (steric hindrance), thereby inhibiting the hydrolytic action of the enzyme.

Lignocaine is available as a 4% solution in multidose containers and also in single-dose disposable units in combination with 0.25% fluorescein for use in tonometry.

There are a large number of other local anaesthetics, but the last four discussed are the only ones available to the ophthalmic optician.

7

Staining agents

Staining agents are used by the ophthalmic optician to determine the accuracy of fit of hard contact lenses or as diagnostic aids. They may be used to reveal damage to the cornea and conjunctiva or to detect conditions of the epithelial cells which may influence contact lens tolerance.

Fluorescein

Types of preparation

Fluorescein as its sodium salt is available in three types of ophthalmic preparation.

Sterile fluorescein papers

These are prepared from filter paper soaked with a 20% solution. The papers are then packed and autoclaved to provide a sterile disposable unit. If the eye is moist sufficient fluorescein may be released by placing the strip in the conjunctival sac for a few seconds. A method which has been recommended to provide a greater volume of fluorescein solution is to crease the strip longitudinally (before opening the sterile pack) remove the strip and place a drop of sterile water or isotonic saline in the crease. The fluorescein solution so prepared can be transferred to the eye.

Sterile single-dose disposable units

These contain 1 or 2% fluorescein sodium without a preservative. They may be used to replace the multidose preparations when larger volumes of fluorescein than are provided by the papers are required. A preparation containing 0.25% fluorescein with 4% lignocaine is also available for use in tonometry.

Eye drops

Fluorescein eye drops, BPC, 1973	
Fluorescein sodium	up to 2%
Phenyl mercuric nitrate/acetate	0.002%
Purified water	to 100%

Multidose containers of fluorescein are very liable to contamination with bacteria including *Pseudonomas aeruginosa.* This organism can produce corneal destruction and is likely to be resistant to a wide range of antibiotics making treatment difficult. Many of the common preservatives (benzalkonium chloride and chlorbutol) are inactivated and chlorhexidine acetate is incompatible with fluorescein. Because of this serious disadvantage the disposable sterile units are to be preferred.

Mode of action

Fluorescein is a water-soluble compound which appears yellow in colour but exhibits a green fluorescence in alkaline conditions. Because of its poor lipid solubility it will not penetrate intact cell membranes. If there is damage to the cells of the epithelial layer fluorescein will gain access to Bowman's membrane, the stroma and even the aqueous. Thus, the stain is taken up by damaged tissue which can be visualized as a green fluorescence in contrast to the yellowish-green appearance of the dye on the undamaged surfaces. If the eye is irrigated with normal saline to remove excess fluorescein the damaged areas are more easily visualized. As the tissue regenerates the colour disappears. Fluorescein drops are non-toxic, locally or systemically, and are non-irritant, even in the presence of damaged tissue.

Uses

Detection of corneal damage

Fluorescein may be used to demonstrate the integrity of corneal epithelium and damage to the conjunctiva.

Contact lens fitting

Fluorescein can be used to identify faults in the fitting of hard contact lenses; the dimensions of the spaces between the lens and the cornea are determined by the amount of colour present. If viewed under blue light, areas of contact between the lens and cornea will be seen as dull purple patches. Any superficial corneal damage during fitting will also be detected.

Fluorescein should not be used with soft contact lens as it is absorbed by the lens material. A high molecular weight fluorescein, fluorexon, has been produced in an attempt to overcome this problem. However the fluorescence, even with 4% solutions, is not always considered to be adequate and the molecule enters some types of lens material, making its use limited.

Detection of foreign bodies

Foreign bodies which are present on the cornea may be more easily detected if fluorescein and a blue light are used. The foreign body is surrounded by a ring of stain which appears green.

Patency of lacrimal ducts

If it is suspected that the naso-lacrimal ducts are blocked fluorescein is a useful diagnostic aid. Fluorescein drops are instilled and if the lacrimal ducts are patent it may be detected in the nasal and oral secretions.

Applanation tonometry

Fluorescein is used in applanation tonometry. The cornea is anaesthetized with local anaesthetic drops and fluorescein applied from sterile papers or combined lignocaine–fluorescein solution used.

Fluorescein is also used by ophthalmologists for fundus photography and studies of aqueous flow.

Rose bengal

This is a brownish-red solid soluble, 1 part in 4 parts water. It is available as a 1% solution in sterile single-dose disposable units.

Uses

Rose bengal stains degenerate cells and their nuclei, and may thus be used to locate degenerate tissue in the sclera and conjunctiva. The presence of facial skin conditions (for example, various forms of dermatitis) may indicate that the cornea and conjunctiva are also affected. Rose bengal may help to diagnose whether or not the ocular tissues are involved. Such tissue would be stained a red colour not removed by irrigation with saline. Care should be taken if local anaesthetic solutions have been used previously as a false positive may be obtained. There is likely to be more irritation after instillation of rose bengal than fluorescein. Particular care should be taken during instillation to avoid overspill onto lids and facial tissues as the staining is relatively persistent.

The use of rose bengal after the wearing of a contact lens will indicate any areas where the lens presses on the cornea and may indicate errors in fitting.

Rose bengal will also stain mucous deposits. In order, therefore, to differentiate between degenerate tissue and mucous, a second stain may be required. Alcian blue, 1%, which preferentially stains mucous may be used for this purpose. This stain should not be used if the cornea is deeply eroded because of prolonged discoloration of exposed connective tissue. Mucous deposits may usually be more simply differentiated by their shape and their movement or removal by irrigation.

8

Antimicrobial agents

The sulphonamides

These are synthetic compounds which have a structural similarity to para-aminobenzoic acid (PABA).

Mode of action

Some bacteria require PABA for growth and multiplication. These bacteria use the PABA to synthesize the vitamin folic acid. When a bacterial cell is exposed to sulphonamides the synthesis of folic acid is blocked due to the competition between the sulphonamide and PABA for the enzyme folic acid synthetase. The similarities between PABA and sulphonamides are shown in *Figure 8.1.* Human cells do not synthesize folic acid but obtain their supply by absorption from the gastrointestinal tract. Therefore, human cells are not affected by sulphonamides.

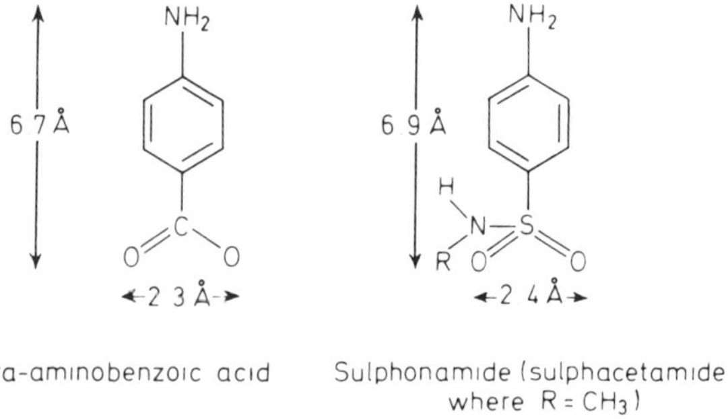

6.7 Å　NH_2　2.3 Å

Para-aminobenzoic acid

6.9 Å　NH_2　2.4 Å

Sulphonamide (sulphacetamide where $R = CH_3$)

Figure 8.1. The molecular structure of para-aminobenzoic acid compared with that of sulphonamides

Sulphonamides have a relatively broad spectrum, acting against Gram-positive as well as Gram-negative bacteria. In the concentrations used they are usually bacteriostatic (that is, they inhibit growth and multiplication), in higher concentrations they may be bactericidal.

Preparations for ophthalmic use

Sulphacetamide sodium is the sulphonamide most widely used by the ophthalmic optician, although a preparation containing mafenide is available. Sulphacetamide sodium is soluble in water giving solutions which are less alkaline than the solutions of most other sulphonamides.

Sulphacetamide drops are available in concentrations of 10, 20 and 30% in multidose containers and 10% in single-dose containers; there is also a preparation containing sulphacetamide 5% and zinc sulphate 0.1%. The eye ointment is available in concentrations of 2.5, 6 and 10%. Mafenide is available as a 5% solution in multidose containers.

Uses

Sulphacetamide drops in a concentration of 10% may be used prophylactically after contact lens work, removal of a foreign body or in other situations in which corneal abrasions are likely to have occurred. The value of this practice is questionable because of the short duration of the bacteriostatic action after instillation. Even the 10% drops are hypertonic and may cause stinging. Although drops of a higher concentration are available, these are usually reserved for the treatment of corneal, conjunctival and lid infections. The use of these drops would be therapeutic rather than prophylactic. The ointment causes less discomfort than the drops and remains longer in the eye. There is a theoretical possibility that some local anaesthetics (for example amethocaine), which are metabolized to PABA or related compounds, may interfere with the activity of sulphacetamide. Thus, if a local anaesthetic has been instilled, it may be wiser to use an alternative non-sulphonamide antimicrobial agent.

Propamidine and dibromopropamidine isethionate

These antibacterial agents are bacteriostatic, bactericidal and also fungistatic. The dibromo compound is available as an eye ointment containing 0.15%. It has a wide range of activity, including an action against Gram-negative bacilli and some strains of *Pseudomonas aeruginosa*. Propamidine is available in isotonic solution at a concentration of 0.1%. It also has a wide range of activity but is not active against *Pseudomonas aeruginosa*.

Antibiotics

The early antibiotics were substances which were elaborated by microorganisms and possessed either bacteriostatic or bactericidal activity. Many of the present antibiotics are synthetic derivatives of these substances. The sale and supply of antibiotics is controlled by law and these drugs are not generally available to the optician, with the exception of framycetin.

Framycetin

This is a broad spectrum antibiotic showing activity against Gram-negative and Gram-positive bacteria. Like other members of the aminoglycoside group of antibiotics, such as neomycin, it is oto- and nephrotoxic and is restricted to topical use. It is available as 0.5% ointment and drops.

Medically prescribed antibiotics

A number of other antibiotics are formulated in preparations for the eye and are frequently prescribed by the medical profession. In general, antibiotics to be applied topically are broad spectrum and should preferably be ones which are not commonly used for systemic infections (because of the possibility of inducing the development of resistant organisms). Those antibiotics which are known to be highly active as allergens, for example penicillins, should also be avoided. Commonly used antibiotics include chloramphenicol, neomycin, gentamycin and tetracyclines.

Chloramphenicol

This has a broad spectrum of activity and is often the drug of choice when the specific cause of the infection is not known. It is available as 0.5% drops (including single-dose) and 1% ointment. Used systemically, chloramphenicol carries the risk of aplastic anaemia and its use is normally restricted to a few life-threatening infections.

Neomycin and gentamicin

Neomycin is similar in activity and potential toxicity to framycetin. It is available in eye drops (including single-dose) and ointment at 0.5% and also in combination with other antibiotics, for example polymyxin B (active against gram-negative organisms) and gramicidin (active against gram-positive). Gentamicin is very similar and is effective against infections caused by *Pseudomonas aeruginosa*.

Tetracyclines

These are broad-spectrum antibiotics with a bacteriostatic action against a wide range of Gram-negative and Gram-positive organisms. They are, however, ineffective against *Pseudomonas aeruginosa*. Chlortetracycline is available as a 1% ointment, tetracycline as 1% ointment or oily drops and oxytetracycline as an ointment with polymyxin B.

The penicillins

The penicillins are of limited use in the treatment of ocular infections. In general, their effectiveness is restricted to Gram-positive bacteria and there is a high incidence of hypersensitivity reactions. Low stability in aqueous solutions and poor absorption give low concentrations, which are dangerous in terms of resistance development. The exceptions which find use in the treatment of ocular infections are benzylpenicillin in gonococcal infections in the newborn and ticarcillin, piperacillin and related compounds (systemically) in the treatment of *Pseudomonas* infections.

9

Solutions used in contact lens work

Some of the solutions discussed will be used only by the contact lens practitioner during fitting, others will also be used by the patient in routine handling and care of the lenses. This latter group will be considered first.

There are now a number of different lenses made of different materials and therefore with somewhat different requirements in terms of care. The care of the two major groups of lenses, namely hard and soft, will be discussed separately. There are, however, some points regarding care which are common to both.

The main requirements of the solutions used in the care of contact lenses are to ensure that on insertion the lens is clean, free from microbial contamination and, in the case of hard lenses, wettable. These requirements have given rise to a considerable number of preparations aiming to perform one, or in some cases all, of these functions. Because of the different nature of the lens materials the problems encountered and the requirements of the solutions varies. Solutions produced for one type of lens should not be used with the other type.

Solutions for use with hard lenses

Three functions are required of solutions used with hard contact lenses: cleaning, removal of bacterial contamination and hydration (wetting) of the lens surface.

Cleaning

During wear and handling lenses may accumulate deposits of protein (from the tears), fats (from the meibomian glands), inorganic materials (from the tears) and organic materials (from handling, nicotine and cosmetics). If these materials are not removed they may collect in the micropores of the lens, thereby reducing the clarity and optical performance of the lens. Allowing the materials to dry on the lens may result in the formation of a scum which is difficult to remove.

The most important constituent of cleaning solutions is of course the cleaning agent itself. This is usually a surface active agent, non-ionic or amphoteric, which is able to emulsify or solubilize debris, regardless of its nature. In addition, the solution will contain an antibacterial compound, not primarily to deal with bacterial contamination on the lens, although it may help in this, but to deal with chance contamination of the solution itself once the container has been opened. The compounds used are normally from the same range used as preservatives in eye drops, namely benzalkonium chloride, chlorhexidine gluconate, thiomersal and phenylmercuric nitrate. These preservatives are generally present in lower concentrations than in eye drops and their effectiveness has been criticized in laboratory conducted bacterial challenges.

Benzalkonium and chlorhexidine are surface active agents and exert their antibacterial effect by causing changes in cell-wall permeability, thus allowing leakage of vital cell constituents. Thiomersal and phenylmercuric nitrate probably exert their effect by combining with thiol (–SH) groups in enzymes, thus inhibiting their activity and interfering with metabolism in the cells.

Many solutions also contain a chelating agent, ethylene diamine tetra acetic acid (EDTA). This will complex with metal ions in the cells (making the metals unavailable for use by the cells in metabolism) or with metal ions in the cell wall (affecting its structural integrity). Either of these actions may enhance the activity of the main antibacterial compound. The solutions may further contain sodium chloride and buffering agents to adjust the tonicity and pH to values approaching that of tears. These properties may not always be adjusted as the solutions are not intended for introduction into the eye. However, the pH may be important as the cleaning agents are usually most effective around pH 7.4 and the stability of the preservatives may be pH dependent; both chlorhexidine and thiomersal are less stable at higher pH values.

Removal of bacterial contamination

When the lens is removed it will be contaminated with bacteria from the wearer's conjunctival sac. Further contamination may occur during handling. Much of this contamination will be removed during cleaning and subsequent rinsing before the lenses are stored, usually overnight, in solution. The function of this storage solution is twofold: to remove the remaining bacterial contamination and to maintain the hydrated state of the lens achieved during wear. These disinfecting/soaking solutions therefore contain as their major constituent an antibacterial compound, again normally from the group used in the preservation of eye drops, namely benzalkonium chloride, chlorhexidine gluconate, phenylmercuric nitrate and thiomersal. Concentrations used are usually lower than those used in the preservation of eye drops (*Table 9.1*), although higher than those present in cleaning or wetting solutions where the agent has to deal only with chance contamination. As already mentioned with cleaning solutions,

TABLE 9.1. Comparison of percentage concentration of antibacterial agents in disinfecting solutions and eye drops

Antibacterial agent	*Disinfecting solution (% concentration)*	*Eye drops (% concentration)*
Benzalkonium chloride	0.001–0.01	0.01
Chlorhexidine	0.002–0.0006	0.01
Phenylmercuric nitrate	0.001	0.002
Thiomersal	0.001–0.004	0.01

doubts have been expressed about antibacterial effectiveness of some disinfecting/soaking solutions and it should be stressed that a sufficient period of disinfecting, preferably overnight, should be allowed and that solutions should be changed frequently, preferably daily, to avoid any compromise of the effectiveness of the antibacterial compound by exhaustion or the build-up of bacterial debris.

Many of the solutions also contain EDTA to enhance the antibacterial activity, sodium chloride to adjust the tonicity and buffers to adjust the pH.

Wetting of the lens

Hard contact lenses are made from polymethylmethacrylate and other ester polymers of methacrylic acid (*Figure 9.1a*). These have near-ideal optical properties but due to the large number of methyl groups they are hydrophobic and not easily wettable (*Figure 9.1b*). If the lens cannot be wetted it will be uncomfortable to wear and, being hydrophobic, will encourage the adhesion to its surface of other hydrophobic substances such as fats. It is possible with the use of certain solutions to convert the normally hydrophobic outer

(a)

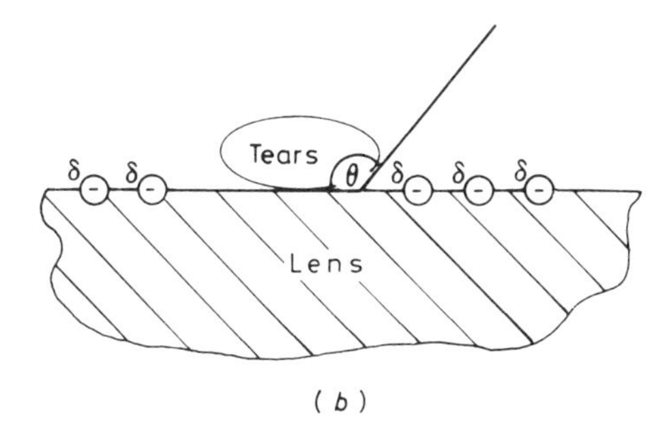

Figure 9.1. Structure and wetting properties of polymethylmethacrylate: (*a*) subunit of the polymer polymethylmethacrylate; (*b*) angle θ for tears at the surface of a polymethylmethacrylate lens. (The high value of angle θ indicates that tears will not spread across the surface of the lens)

layer of the lens to a hydrophilic one. Most wetting solutions contain polyvinyl alcohol as the wetting agent. Polysorbate 80 can also be used but concentrations above 0.5% may inactivate some antibacterial compounds such as benzalkonium chloride and chlorhexidine.

The mechanism by which wetting agents achieve their effect is shown in *Figure 9.2.* At the surface of the lens, solute molecules orientate so that the hydrophilic end of the solute molecule is presented to the aqueous phase and the hydrophobic portion is presented to the hydrophobic plastic; this gives a hydrophilic interface and allows wetting of the plastic by the tears. Many solutions also contain viscosity-increasing compounds such as methylcellulose and derivatives, which render the solution more viscous so that it adheres better to the lens surface.

Wetting solutions contain antibacterial agents to deal with chance contamination during use: benzalkonium chloride, chlorhexidine gluconate, thiomersal and phenylmercuric nitrate are normally used. Concentrations are generally below those used

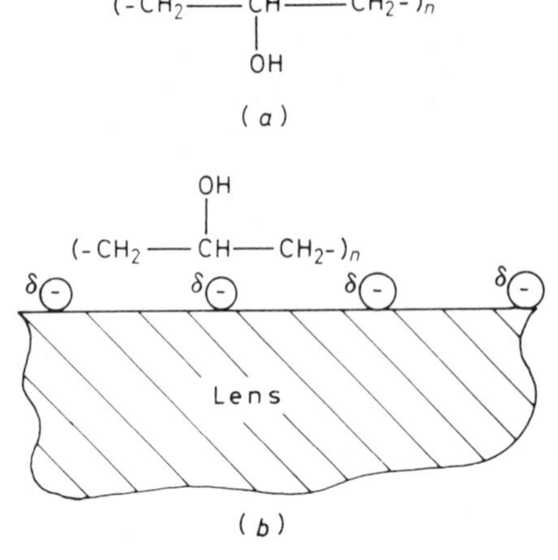

Figure 9.2. Structure of polyvinyl alcohol and its orientation at a lens surface: (*a*) subunit of polyvinyl alcohol; (*b*) orientation of polyvinyl alcohol at the surface of the lens

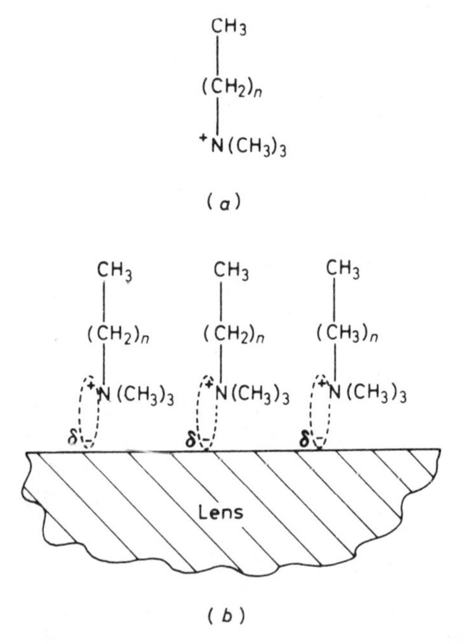

Figure 9.3. Structure of benzalkonium chloride and its orientation at a lens surface: (*a*) subunit of benzalkonium chloride; (*b*) orientation of benzalkonium chloride at the surface of the lens

in eye drops or disinfecting solutions to avoid the possibility of irritation to the eye from prolonged contact. Although benzalkonium chloride is a surface active agent, it does not act as a wetting agent (*Figure 9.3*), it orientates with its hydrophilic portion towards the surface of the lens (due to ionic attraction), thus producing a new hydrophobic surface. This becomes more marked as the concentration of benzalkonium chloride is increased. Many wetting solutions also contain EDTA to enhance antibacterial activity.

As wetting solutions come into contact with the eye, it is more important that tonicity and pH are adjusted for maximum patient comfort. Solutions with a tonicity equivalent to 0.6 to 1.5% sodium chloride can be tolerated by the eye but contact lens wearers may be more sensitive and to avoid discomfort due to fluid movements at the cornea, solutions are better adjusted with sodium chloride to be equivalent to 0.9 to 1.1%. Again, to reduce patient discomfort and avoid excessive tear production, the pH should be buffered to approximately 7.4.

The use of wetting solutions, as well as giving a wettable and therefore more comfortable lens, will also cushion the lens at the corneal surface and act as an aid to cleaning after removal.

There are solutions available which aim to combine more than one function or indeed all functions of cleaning, disinfecting, soaking and wetting. While such solutions may be simpler for the patient, it is possible that there will have to be compromises between different functions. For example, the high viscosity of solutions with a wetting function may not be ideal for antibacterial activity. Any mixed-function solution which is to be used on the lens for insertion in the eye requires a lower concentration of antibacterial compound to avoid possible irritation.

Solutions for use with soft lenses

The requirements from solutions used with soft contact lenses are they they should render the lens clean and free from bacterial contamination. The materials used in the manufacture of soft lenses are based on polyhydroxyethyl methacrylate (polyHEMA) and are themselves hydrophilic, due to the presence of hydroxyl groups, so that no wetting solutions are required.

Cleaning

Soft lenses are liable to the accumulation during wear of the same type of debris as hard lenses. The deposits in question are lipids from the meibomian glands, protein and inorganic deposits, particularly calcium, from the lacrimal glands. The accumulation of these deposits may lead to reduced optical performance from the lens, poor wetting of the lens, binding of cationic preservatives such as chlorhexidine and allergic responses due especially to the build-up of protein material. The situation with regard to protein will be worsened if the lenses are boiled to disinfect them, as this will result in denaturation of the protein on the lens surface. It is therefore essential that these contaminants, especially the protein, are removed. If they are not, the performance of the lens, and its life, are likely to be reduced and the effectiveness of certain disinfection procedures adversely affected.

More than one approach to the cleaning of soft lenses exists. One is the same as that for hard lenses, namely the use of surface active agents. These will emulsify and solubilize many of the deposits and this process is aided mechanically by rubbing the lens. In some cases the solutions are made hypertonic, movement of water from the lens assisting the removal of absorbed particles. Changing pH may also have an influence; alkaline pH tends to aid removal of protein while acid pH may be used to increase stress at the lens surface to help crack deposits located there. Solutions also contain an antibacterial agent and EDTA to deal with chance contamination of the solution itself. Because of the ability of the soft lens material to absorb and bind many substances the use of antibacterial compounds has problems as will be discussed below.

Although the regular use of surfactant cleaners will remove many contaminants and make subsequent disinfection of the lens easier, it will not remove denatured protein. This may give rise to an allergic response and encourage the absorption of preservatives from solutions. Such protein build-up may be removed by the periodic use of enzyme cleaners. These usually contain, in tablet form, proteolytic enzymes which are capable of hydrolysing proteins to yield smaller peptide fragments which are more easily removed from the lens. Papain (extracted from paw paw) is widely used but other proteolytic enzymes are available. Pronase is claimed to have a broader range of activity and to be able to degrade mucin. Also available in tablet form is lipase which breaks down triglycerides (fats and oils) to give glycerol and fatty acids which are more easily removed than the original molecules. Enzyme tablets may also contain buffers and EDTA.

A third approach to the cleaning of soft contact lenses is the use of oxidizing agents. This method of cleaning, which is generally reserved for the practitioner rather than the patient, is used to deal with deposits which have built up despite daily surfactant and periodic enzyme cleaning. The cleaning depends on the presence of oxidizing agents or the generation of oxygen at an elevated temperature. In both cases any deposits, regardless of their nature, will be removed by oxidation. After cleaning with oxidizing agents the lens must be thoroughly rinsed to remove all traces of this agent. The actual details concerning the use of both these agents and the enzyme cleaners will vary with individual preparations and the manufacturer's instructions should be carefully adhered to.

Removal of bacterial contamination

Soft lenses, like hard ones, become contaminated with bacteria during wear and it is necessary that this contamination is removed during the non-wear or storage period. Because of the nature of the lens material and its ability to concentrate and absorb chemicals from solution, the preparations used for disinfecting hard lenses are not suitable.

The original approach to the disinfection of soft lenses was to boil them. While this method is effective, there are some disadvantages in that spores are not killed and the useful life of the lens may be reduced by this treatment. Of particular importance is the fact that any protein remaining on the lens – despite prior cleaning – will be denatured, more difficult to remove and may precipitate allergic responses. Patients may also find the procedure inconvenient and be tempted to be less than rigorous in adhering to a daily disinfecting routine. Because of these disadvantages attention was turned to chemical disinfection.

The problem with chemical disinfection of soft lenses is that the commonly used antimicrobial agents, benzalkonium chloride, chlorhexidine and thiomersal are all bound to or absorbed by the lens material to a greater or lesser extent. While these compounds cause no ocular problems in the concentrations normally used, their accumulation by and binding to the lens may bring higher concentrations into contact with the eye for longer periods, resulting in discomfort and possible allergic responses. Benzalkonium is the most highly bound and is not used in soft contact lens solutions. Binding appears to be less marked with the other two and many disinfecting solutions contain a combination of both chlorhexidine (0.005–0.005%) and thiomersal (0.001%). The combination is used for two reasons. First, chlorhexidine has excellent activity against bacteria, but is only mildly effective against fungal contamination; thiomersal possesses the reverse activities showing excellent activity against fungi. The second reason for the combination is the apparent ability of chlorhexidine to render the lens surface hydrophobic (possibly due to its cationic nature causing it to orientate in a similar manner to benzalkonium). Low concentrations of thiomersal prevent this occurring. This combination appears to possess an effective disinfecting action on soft contact lenses. A number of patients report irritation from the use of solutions containing this combination or chlorhexidine alone and, while in some cases it is undoubtedly a true response to the antibacterial agent, it may in some cases be an allergic response to protein build-up. EDTA may also be present to enhance antibacterial activity as in other solutions. Ideally the pH and tonicity should be similar to that of the tears, particularly if the lenses are to remain in the solution for any length of time.

An alternative to the use of chlorhexidine, with or without thiomersal, is the use of hydrogen peroxide. There are a number of products with variations in the routine, but the essentials involve soaking the lens in 3% hydrogen peroxide for some time followed by a further soaking to bring about neutralization. This may be achieved by the use of sodium bicarbonate or sodium metabisulphite solution or a platinum catalyst in preserved saline, which aids the decomposition of the hydrogen peroxide to oxygen and water. In patients who experience irritation with the usual preservatives, non-preserved saline may be substituted in the later stages. The antimicrobial action is thought to be one of oxidation following the release of free radicals from the hydrogen peroxide. This will also bring about some cleaning of the lens, although protein deposits may not be completely removed. Whichever hydrogen peroxide system is used, there is considerable handling of the lens and patients must adhere strictly to the routine to ensure that the lens is not inserted into the eye while hydrogen peroxide remains on it. Although considerable discomfort and signs of ocular irritation follow such a procedure, evidence indicates that there is no permanent damage caused to the ocular tissues.

Active iodine generated from an iodophor has an effective bactericidal activity. In this system the already clean lens is placed in a solution containing the iodophor in a polymeric vehicle together with a preserved neutralizing solution. The iodine is converted to iodide, at the same time exerting its antibacterial activity and the solution loses its colour. It has been shown that the time needed for the decolourization exceeds that needed for the antibacterial activity. Thus it can be assumed that if the solution has lost its colour, the disinfection process is complete. The preservatives used, sorbic acid, sodium borate and EDTA are not bound to the lens material so that problems of irritation should not occur.

For lenses which are disinfected using hydrogen peroxide or iodophors, it is necessary to have a soaking solution in which the lenses remain between cleaning/disinfecting and wearing. This is usually normal saline preserved with chlorhexidine, thiomersal and EDTA to deal with chance contamination. It is important that solutions in which soft lenses are to remain for any time should be isotonic to maintain the correct hydration and dimensions.

In addition to the solutions used for cleaning and disinfecting (and storing) soft lenses, there are available rinsing solutions which can be used during disinfection procedures or to wash away cleaning solutions. These include single-unit, non-preserved saline and larger volumes of saline preserved with thiomersal (with or without EDTA) or sorbic acid, sodium borate and EDTA. These preservatives are intended to deal only with chance contamination of the solution, not to exert any disinfecting action on the lens. Some solutions used primarily for disinfection may also be used for rinsing. These, as discussed earlier, generally contain a combination of chlorhexidine, thiomersal and EDTA.

As all soft lens materials are not identical, there may be individual variations concerning the suitability of a particular solution: for example some materials concentrate chlorhexidine to a greater extent or are discoloured by iodine. Such information should be supplied by manufacturers and must be taken into consideration when deciding which solutions to use or recommend.

The remaining solutions are those used by the optician during fitting. They include local anaesthetics (Chapter 6), staining agents (Chapter 7), antibacterial drugs (Chapter 8) and vasoconstrictors (Chapter 10).

10

Decongestants, antihistamines and anti-inflammatory compounds

Decongestants

The drugs which are included in the sympathomimetic mydriatric group may also be used as decongestants. Peripheral blood vessels such as those present in the conjunctiva receive post-ganglionic sympathetic innervation. The activity of these nerves results in vasoconstriction due to the release of noradrenaline and its interaction with α-receptors. Thus, adrenaline, phenylephrine and ephedrine will produce vasoconstriction when applied topically to the eye. The mechanisms by which they exert their effects have been discussed in the section on mydriatics. Compounds which are not used to produce mydriasis, but which are used as ocular decongestants, include naphazoline and xylometazoline hydrochloride. These are potent, directly acting, α-receptor stimulants.

Preparations

Adrenaline solution

This may be prepared from either the tartrate or the hydrochloride salts giving a concentration of 0.1% (1 in 1000) adrenaline. The solution may also contain chlorbutol and chlorocresol as preservatives, sodium metabisulphite as antoxidant and sodium chloride to produce an isotonic solution. The solution should be stored in a cool place and protected from the light.

Phenylephrine and ephedrine

These substances may be used in solutions of the hydrochloride salt to produce vasoconstriction. A satisfactory decongestant action can usually be obtained with lower concentrations than those needed when a mydriatic effect is required. Phenylephrine, for example, is available in a number of preparations at a concentration of 0.12%. Mydriasis would not normally be expected at this concentration.

Naphazoline and xylometazoline

Naphazoline may be used as a decongestant in solutions containing 0.05 or 0.1%. The solution should be adjusted to a suitable pH and tonicity. A preservative such as phenyl mercuric nitrate may be included and the solution should be stored protected from light. Xylometazoline is usually prepared as a 0.05% solution in a formulation which also contains an antihistaminic drug.

Uses of decongestants

These solutions may be used in contact lens work to prevent vasodilatation which can occur during moulding and which would produce irregularities in the mould. The effects of adrenaline, 1 in 1000, usually last for less than one hour and may be followed by reactive hyperaemia. If vasoconstrictor solutions are used as decongestants to blanch reddened conjunctival vessels due to non-specific chronic irritation or allergy, the longer acting sympathomimetic drugs such as phenylephrine may be preferred. Prolonged use of vasoconstrictor agents is not to be recommended because of the danger of masking the symptoms of a more serious condition. With the longer acting drugs, such as naphazoline, the effect may last several hours, sometimes making the rebound dilatation more marked.

Antihistamines

During an allergic reaction, for example, hay fever or hypersensitivity to atropine, it is thought that histamine and certain other substances present in tissues (5-hydroxytryptamine and various kinins) are released. Normally these substances are stored in a bound inert form but when released into the blood stream or extracellular fluid they affect tissues to produce itching, constriction of smooth muscle and vasodilatation. There is also an increased capillary permeability resulting in leakage of capillary contents and oedema. Thus, an allergic condition of the eye would present as a swollen, inflamed conjunctival area with severe itching and discomfort. The ultimate remedy for this condition is the removal of the allergen. Temporary symptomatic relief of the condition may be achieved by the use of a vasoconstrictor (*see* above) and an antihistamine drug. The naturally occurring substance histamine will produce most of the signs of an allergic response when applied directly to the tissues. The antihistamine drugs compete with histamine for receptors in the tissue thereby antagonizing its actions. This antagonism is of the competitive type. The sympathomimetic drugs are not competitive antagonists but are described as 'physiological' antagonists. These drugs do not compete at the histamine receptor but produce vasoconstriction by a different mechanism, thereby reducing or abolishing the histamine vasodilatation.

A preparation available to relieve the symptoms of an allergic condition of the eye contains antazoline sulphate (0.5%) and xylometazoline hydrochloride (0.05%). Benzalkonium chloride may be used as the preservative.

A large number of antihistamine drugs is available but are mostly formulated for oral administration use in the systemic relief of allergic conditions. Antazoline sulphate is chosen for use in the eye because it is less irritant to the eye than most other antihistamines applied topically.

Anti-inflammatory compounds

These drugs include the naturally occurring adrenocortical steroids and synthetic derivatives developed from the naturally occurring compounds. They are able to suppress and prevent all signs of inflammation, for example, redness and swelling, regardless of the initial cause. It is most important to note that steroids only prevent the signs of inflammation, they do not remove the underlying cause. Because of the danger of masking the development of serious systemic conditions the corticosteroids should only be used under medical supervision.

The systemic use of these compounds produces a number of undesirable effects including hypertension, oedema, electrolyte disturbance and hyperglycaemia. When used topically there is little danger of systemic effects.

The two most widely used preparations are hydrocortisone and prednisolone in the form of eye ointments or drops. They may be used by the medical profession for the symptomatic treatment of blepharitis, conjunctivitis and other local inflammatory reactions. There are two main dangers with topical use. First, a rise in intraocular pressure (due to a decreased outflow of aqueous humour) in persons predisposed to such an effect and second, the inadvertent use in the presence of inflammation, of bacterial, fungal or viral origin, allowing the infection to proceed even though the signs of its presence may be suppressed.

There is a second group of anti-inflammatory compounds which includes aspirin; these are known as the non-steroidal anti-inflammatory drugs. Members of this group produce their anti-inflammatory effects by inhibiting the synthesis of prostaglandins which are involved in the later stages of inflammation. Different drugs within the group show differing analgesic and antipyretic activities; these effects are also believed to be related to the inhibition of prostaglandin formation.

One of this group, oxyphenbutazone, is available to the optician as a 10% ointment for use in non-infective inflammatory conditions of the anterior segment.

The release of histamine and other mediators associated with an allergic or inflammatory response may be prevented by the use of sodium cromoglycate, which acts by stabilizing mast cell membranes. It is formulated in a number of preparations, including 2% eye drops and ointment and are used for allergic conjunctivitis. It is not available to the optician and is of doubtful value if given after the response is manifest.

11

Ocular effects of drugs used systemically

In recent years everyone concerned with drug use, including patients, has become increasingly aware that drugs produce effects other than those required of them. Such unwanted effects may involve ocular tissues and function. The problems may arise from drugs used topically: those likely to occur following the use of drugs by the optician, especially closed angle glaucoma associated with mydriatics, have already been mentioned. It is likely that in the case of ocular side-effects following the topical use of drugs by the general practitioner or ophthalmologist, the patient will return directly to them (rather than the optician) as the connection between the problem and the drug should be obvious. Problems from the use of topical drugs, other than those already described, include the tendency for antibiotics to cause allergic responses and allow the overgrowth of nonsusceptible organisms. It is also possible that timolol used for the treatment of glaucoma may possess the potential to induce a syndrome similar to the keratoconjunctivitis sicca and other problems seen with the now-restricted practolol.

More likely to come to the attention of the optician are those ocular side-effects occurring as a result of the systemic use of drugs. The reasons the optician is likely to encounter these are twofold. First, if the patient is aware of visual problems, it is unlikely that the connection between these and systemic medication will be made. Second, the optician sees many patients regularly and is in a position to detect those changes in vision or ocular tissues of which the patient may be as yet unaware. It is essential for the optician to be alert to the possibility that ocular changes or visual disturbances detected on examination or reported by the patient may be due to systemically administered drugs. Such awareness may enable serious permanent damage to tissues and vision to be avoided by prompt referral of the patient, or prevent unnecessary changes being made to a patient's prescription to cope with what may be only temporary drug-induced changes in accommodation. During history-taking the optician should establish whether or not drugs are being taken regularly and if so determine their identity. This latter should not be difficult as all dispensed medicines are now labelled with their name. Some patients, however, may not be able to remember drug names correctly. In such cases they are usually able to inform the optician of the reason for taking the drug(s) and it is useful therefore to possess some source of information which indicates the drugs or groups of drugs used in different conditions. When questioning patients about drug-taking, it should be remembered that some drugs capable of inducing side-effects are available from a pharmacy without a prescription and patients may be using these on a regular basis without thinking of them as medicines and thus neglect to mention them. It may also be important to establish whether the drug-taking is likely to continue for any length of time or whether it is more likely to be a short-term temporary situation. This is of particular importance with those drugs possessing the potential to disturb accommodation.

Should the optician detect any changes in ocular tissues or visual performance the matter should be referred, with details of the findings, to the patient's doctor for consideration. It may be that the doctor is unaware of the problem and that the drug in question can be changed. In some cases it is possible that the advantages of the particular drug are such as to outweigh the side-effects. This will depend on the disease being treated, the benefit to the patient of the drug and the nature of the side-effect, and is a matter for the doctor to decide. It should be pointed out that the term 'side-effect' implies any effect other than that for which the drug is being used and does not of necessity imply toxicity, detriment to the patient or permanent damage. If the matter is to be referred the optician must ensure that this is done without causing undue alarm to the patient with the risk that essential medication is abruptly stopped. The consequences of this could be more dire than the side-effects detected.

Although a vast number of drugs have been cited as implicated in causing side-effects in the eye, the majority of patients will experience no problems whatsoever. The number of ocular problems which may be related to drug use as reported to the Committee on Safety of Medicines in Great Britain is not high and, while accepting that only a percentage of such problems are reported, it can probably be assumed that the ones reported are the more serious ones. As with any side-effects reported, it is not always easy to be certain that the drug cited is in fact responsible, particularly when many patients take several drugs concurrently. All patients do not show side-effects but a number of factors may influence the likelihood of problems arising. For instance, with most drugs the higher the dosage used and the longer the administration is continued the greater the chance of side-effects becoming apparent. This is certainly true in the case of chloroquine retinopathy or steroid-induced cataracts. Age may also play a role, the eye may be predisposed to the side-effect because of ageing; decreased liver and kidney function may allow higher concentrations of the drug to stay in the body for a longer time thus allowing higher concentrations to be achieved in the eye. The incidence of a side-effect may be genetically determined, as with steroid-induced rises in intraocular pressure. Genetically determined rates of drug metabolism, as with isoniazid which causes peripheral neuropathy including optic neuritis, can also influence the incidence of side-effects. Some 60% of the European population have a slow rate of metabolism of isoniazid and are more likely to experience problems.

There exists a bewildering array of ocular side-effects and drugs reported to cause them. The effects range from minor disturbances in pupil size and accommodation, which are fully reversible on withdrawal of the drug, through to permanent damage to ocular tissues with accompanying deterioration in visual function. The drugs implicated cover all therapeutic groups and while in some cases the explanation for the side-effect is obvious (or implied) from the known pharmacology of the drug, in others it is not. It is possible that the optician may encounter effects apparently attributable to drugs which are not well documented due to a low incidence of occurrence or because the drug has only recently been introduced. Opticians are encouraged to report suspected ocular side-effects of drugs to the British College of Ophthalmic Opticians (in addition to referring the matter to the patient's doctor).

It is not possible to provide an exhaustive list of all the drugs which have been implicated in causing ocular side-effects, so an attempt has been made to confine the discussion mainly to those drugs and effects which are most well known, most common or most serious. Further information is available in specialized texts dealing with this problem or, in answer to specific queries, from drug information services.

A number of approaches to the discussion of drugs with ocular side-effects are found in texts. Drugs may be discussed in pharmacological or therapeutic groups or in alphabetical order. As the optician will see changes, if any are present, structure by structure, another approach is to consider the effects produced on each structure and some of the drugs responsible. Using this approach the same drug may appear more than once.

Conjunctiva

High doses or prolonged therapy with phenothiazines used in psychiatric medicine may result in pigmentation of the conjunctiva. This has most often been reported with chlorpromazine which is probably the most widely used of the group and in some patients may be associated with a photosensitivity reaction which results in pigmentation of the skin.

It is unlikely now that the optician will encounter the oculomucocutaneous syndrome induced by practolol as the drug is restricted to hospital use. However, any conjunctival abnormalities in patients being treated with beta-blocking compounds are grounds for referral.

The sulphonamides may induce allergic responses of the conjunctiva including the Stevens–Johnson syndrome. This latter is rare and usually associated with the long acting members of the group.

Cornea

The use of chloroquine in the treatment of rheumatoid arthritis can cause the appearance of a whorl-like pattern of deposition within the cornea. This normally does not affect vision, is reversible on withdrawal of the drug and is believed to be due to the formation of complexes with cellular phospholipids. Deposition is less likely when the drug is used for the treatment or prophylaxis of malaria, as the duration of treatment is shorter in one case and the dose used lower in the other. A similar pattern of deposition is seen with the phenothiazines such as chlorpromazine, and in almost all patients treated with amiodarone for cardiac arrhythmias. In this case the appearance of the deposits is so common as to almost be an indication that the drug is being taken correctly. Indomethacin, an antipyretic analgesic with marked anti-inflammatory activity which is in widespread use in rheumatoid conditions, has also been implicated in causing corneal deposits and it remains to be seen if the same will apply to other members of the same group. Oxyphenbutazone and phenylbutazone which commonly caused ocular side-effects have recently been either withdrawn from systemic use (oxyphenbutazone) or severely restricted in use (phenylbutazone) so that the optician is now less likely to encounter problems from these drugs.

A second problem with the cornea is that of oedema which may influence the wearing of contact lenses. This is associated with the use of corticosteroids in asthma or rheumatoid conditions and oestrogens in the contraceptive pill. As both groups of drugs influence electrolyte and water metabolism in the body this is not entirely unexpected. The same problem may arise with the lens resulting in blurred vision.

Iris and ciliary muscles

Not unexpectedly, those drugs which are muscarinic antagonists (that is, atropine-like) have the potential when used systemically to cause relaxation of the ciliary and sphincter muscles. The patient may thus experience difficulties with accommodation and be disturbed by bright lights because of the pupil dilatation and loss of the light reflex. More seriously, the dilatation of the pupil may lead to angle block and a rise in intraocular pressure in predisposed patients. While such actions may be anticipated from muscarinic antagonists it is not always obvious that the drugs the patient describes are in fact muscarinic antagonists. Drugs are used primarily as muscarinic antagonists in peptic ulcer, hypermotility of the gastrointestinal tract, Parkinson's disease and travel sickness, but many other drugs possess muscarinic blocking activity to varying degrees as secondary actions. These include the neuroleptics such as the phenothiazines and haloperidol, the tricyclic antidepressants such as amitriptyline and the antihistamine compounds used in allergic conditions.

Mydriasis with the risk of angle block may also be produced by the use of sympathomimetic drugs, which cause the dilator muscle to contract. Such drugs may be encountered in various cough and cold remedies including those available without a prescription.

Lens

Probably the most well-documented side-effect on the lens is the tendency for the corticosteroids used over prolonged periods, particularly in rheumatoid conditions, to cause the formation of posterior subcapsular cataracts. This may be a reflection of interference by the steroids with the metabolism of the lens. Physiologically, steroids are involved in all aspects of metabolism: fat, protein, carbohydrate and electrolyte but the presence of high concentrations can lead to disturbances.

Phenothiazines can produce granular deposits in the lens which lead to stellate or anterior polar cataracts.

The possibility of cataracts from the use of the long-acting anticholinesterases has already been mentioned.

Retina

Retinal pigmentary changes associated with a narrowing of retinal arterioles and a reduction in visual acuity is a possibility with the phenothiazines, particularly thioridazine. The changes are normally rare and result from high dosage and prolonged treatment and may progress after the drug is withdrawn because of binding to the retinal pigment.

Chloroquine induces retinal and macular changes (Bull's Eye macula), narrowing of retinal vessels and deterioration in vision. The explanation for the effects is not clear. If the drug is withdrawn early, the maculopathy is reversible. These changes are normally only associated with high dosage over prolonged periods and, as the problem is well known, patients being treated with chloroquine are likely to be referred for regular ophthalmological examination.

Many other drugs have been suggested as causing retinal damage but the link is not always clear. Obviously any abnormal retinal findings require that the patient be referred.

Retinal oedema has been reported in connection with the use of both corticosteroids and the thiazide diuretics. This may be due to their influence on electrolyte and fluid metabolism. Papilloedema secondary to pseudotumor cerebri may follow abrupt cessation of steroid administration.

Optic nerve

Most of the drugs used in the treatment of tuberculosis, namely ethambutol, isoniazid, rifampicin and streptomycin have been implicated in causing optic atrophy, optic or retrobulbar neuritis and disturbances in colour vision. This latter is usually manifested as a red–green defect, although other difficulties may be experienced. Ethambutol is the drug most likely to cause problems but, by keeping the dose to 15 mg per kg, these can usually be avoided. The optic neuritis induced by isoniazid is one aspect of the peripheral neuropathy this drug causes. Concurrent administration of pyridoxine prevents this. In most patients treatment of tuberculosis is prolonged and involves a combination of drugs. As with other drugs where ocular side-effects are well documented it is likely that such patients will have their visual function monitored regularly.

Colour vision

Disturbances of colour vision are common with nalidixic acid which is used for urinary tract infections. Objects may have a green, yellow, blue or violet tone, there may also be flashing lights or a glare phenomenon. Similar problems of glare phenomenon and disturbances in colour vision may be experienced with the antiepileptic drug troxidone but this is now little used. Digoxin and other cardiac glycosides used in heart failure and certain arrhythmias cause a blue–yellow defect in colour vision associated with glare phenomena, flickering vision, haloes round lights and other disturbances. These effects may be central in origin or due to a selective effect on cone cells. As digoxin has only a very narrow therapeutic range and the ocular problems generally precede the more serious cardiac disturbances, it is essential that these signs are recognized so that the patient may be referred for dosage adjustment.

Ocular movements

Many drugs have been reported to cause nystagmus. These drugs are often central nervous system depressants including the barbiturates and other hypnotics, the anxiolytics such as diazepam and anticonvulsants such as phenytoin. The problem may thus be central in origin, rather than a direct effect on the extraocular muscles. With streptomycin the nystagmus may have a vestibular origin as the drug affects the eighth cranial nerve.

Blurred vision

Probably the most common ocular side-effect reported by patients is that described simply as blurred vision (which is found not to be due to relaxation of the ciliary muscle). This has been reported for a number of diuretics, where the explanation may be interference with ion and fluid exchanges between the aqueous and the lens; for the anxiolytics where the explanation may be central interference with extraocular muscle movements as these drugs possess central muscle relaxant properties; for the oral hypoglycaemics where the effect may be due to hypoglycaemia and to an almost infinite number of other drugs where there is as yet no explanation for the disturbance. Fortunately, these effects are generally transient and reversible on cessation of the drug.

12

First-aid and emergency measures used by the ophthalmic optician

As opticians are not allowed to treat conditions of the eye, first aid by the optician should be only what the words 'first aid' imply. Any measures taken should be sufficient to minimize any harmful processes and reduce the discomfort of the patient then, if necessary, the patient should be referred to a doctor or the accident unit of a hospital.

The problems commonly met by the optician include the following.

Foreign bodies

If foreign bodies are easily removed, it is permitted to do so using a catgut loop or other suitable implement. A local anaesthetic instillation may make this task easier. A sterile fluorescein preparation to check for the presence of abrasions, and the prophylactic application of an antibacterial substance may be required. If the foreign body has penetrated the ocular tissues referral is essential. When a local anaesthetic is used care must be taken that the loss of the blink reflex does not expose the patient to the danger of further damage due to undetected foreign bodies. Therefore, the patient may be detained until the blink reflex has returned or if this is not possible, the eye could be covered with a pad. If both eyes have received local anaesthetic instillations, some protection may be afforded by the instillation of lubricant drops containing polyvinyl alcohol or hypromellose, or castor oil, which is available in single-dose disposable units.

Other injuries

If the ocular tissues have received severe cuts, bruises or similar injuries, it is necessary to refer the patient to a doctor.

Harmful chemical agents

Lime (quick lime, calcium oxide)

In the case of lime burns, it is necessary to wash the eye as quickly as possible with water or sterile isotonic saline using a local anaesthetic to aid this if necessary. Some authorities recommend that washing should be continued for as long as thirty minutes. After this, any adherent particles should be removed. If available a 10% solution of ammonium tartrate may be used which will aid the removal of the particles. The chelating agent sodium edetate may be substituted for the ammonium tartrate.

Sulphuric and other mineral acids

In the case of acid burns to the eye immediate and prolonged washing with water or sterile isotonic saline is necessary. If available, sodium bicarbonate eye lotion may be applied but any advantage over water is minimal.

Caustic soda and other caustic solutions

In the case of burns with alkaline solutions, irrigation with a large volume of water or sterile isotonic saline is necessary. If available, citric acid eye lotion or a saturated boric acid solution may be used.

Ammonia

Ammonia burns of the eye should be treated by washing with water or sterile isotonic saline. The washing may be followed by the instillation of liquid paraffin drops. In severe cases, the anterior chamber may need draining because of the destructive effects of absorbed ammonia on the vitreous humour and retina. This procedure obviously is not first aid and referral is essential in such severe cases.

In most emergencies due to chemical injury, the main safeguard is a mass irrigation of the eye which produces dilution of the offending agent, or physical removal in the case of non-water miscible liquids. In all cases where damage to the ocular tissues is suspected or where there is a risk of infection, medical attention should be obtained immediately following the first-aid measures.

Acute close angle glaucoma

This may be precipitated by the optician by the use of a mydriatic or a patient may present with an attack. If any first aid is to be rendered, it is essential that the optician recognizes the attack for what it is and is able to differentiate it from iritis. An acute glaucoma attack is characterized by its sudden onset and intense pain. The eyeball is stony hard, there is a faded iris, a cloudy oedematous cornea, a congestive hyperaemia in the conjunctival vessels and a dilated, irregular pupil which does not respond to light. The condition may be accompanied by nausea and vomiting. The patient should be referred immediately. Although it is possible to instil pilocarpine (2 drops of 1% in each eye every 10 minutes for 30 minutes), some ophthalmologists prefer to see the patient untreated. It is wise therefore for the optician to determine the preferred approach in the area and to take this and any delay in the patient receiving medical attention into consideration when deciding whether or not to instil pilocarpine.

Iritis

This condition is characterized by a gradual onset. The intraocular pressure is usually normal, the iris is faded, precipitates are seen in the cornea and the pupil is constricted and sluggish. Again, the patient should be referred to a doctor or a hospital. A mydriatic may be instilled to reduce the risk of posterior synechiae forming but the possibility of inducing closed angle glaucoma must be considered.

Patient collapse

It may happen that a patient faints or collapses while in the practice and the optician should ensure that the measures to be taken in dealing with such problems are known. Fainting is the result of a sudden fall in blood pressure which may be due to the activation of the sympathetic system. The patient loses consciousness but respiration and the pulse are maintained. All that is normally required is for any tight clothing round the neck to be loosened (the optician should avoid being alone with the patient in this situation) and for the patient's head to be placed between his or her knees. No attempt should be made to force liquids on the unconscious patient. Once consciousness is regained the patient may be fully revived by fresh air and a rest.

Sudden collapse may be due to a number of causes and there may be cardiac arrest followed by loss of respiratory function or initial loss of respiratory function followed by cardiac arrest. Whatever the cause, it is essential that attempts are made to restore and maintain both respiratory and cardiac function until the arrival of medical assistance. In the absence of cardiac function (manifested by the lack of a pulse) cardiac compression at the rate of some 60 per minute should be applied in an attempt to maintain the circulation. If respiration has ceased, mouth to mouth resuscitation should be applied at the rate of around 12 per minute, after first making sure that the airways are clear. In some cases both respiration and cardiac function may be absent and attempts should be made to restore both, either involving two people concurrently or, if only one person is available, cardiac compression should be alternated with mouth to mouth resuscitation. In all cases the attempts should continue until the functions are restored or medical assistance arrives.

Records

In all cases of first aid, the optician should keep records with details of the patient, the problem, any treatment given and a note of the patient's referral to hospital (if applicable) or advice given to contact a doctor or hospital should there be any pain, discomfort or deterioration in vision. Other than questions that are necessary to determine the problem details concerning the patient, the patient's doctor should be obtained after the immediate problem has received attention.

13

Formulation of eye preparations

The preparations to be considered are eye drops, eye ointments and eye lotions.

Eye drops

When formulated and prepared, eye drops should contain the drug in such a form that it is well absorbed from the corneal surface and stable in the formulation. The solution should be sterile (and have some means of maintaining sterility). The ingredients of the eye drops should be compatible with each other and should not interact with the container or fastener. Attention may also be given during formulation to the tonicity of the solution. Isotonicity is preferable although not essential. The pH of the solution should be as near neutral as possible.

Absorption of the drug

This has been discussed at some length in Chapter 2 and the Appendix. Important factors for absorption are the pH of the solution and the pKa of the drug. If it is necessary to use a pH markedly different from neutral, the drops may not be well tolerated by the patient. There may be stinging, irritation and probably loss of solution due to excessive tear production. Some authors suggest that most solutions between pH 3.5 and 10.5 can be tolerated. It is possible to aid the penetration of drugs by including in the formulation a wetting agent such as benzalkonium chloride. Wetting agents aid penetration by lowering surface tension and increasing the permeability of the membrane.

Stability

Since drops are not usually freshly prepared, they must be formulated so that the active drug has a reasonable shelf life. Several factors may increase the rate of deterioration, for example, oxidation, pH, light, heat and hydrolysis.

Oxidation

A number of drops, for example, physostigmine are oxidized to yield products which may be less active and irritant. The oxidation process may be reduced by including in the formulation an antoxidant. Antioxidants are compounds which are preferentially oxidized. The antoxidant commonly used in eye preparations is sodium metabisulphite.

pH

The stability of drugs will vary depending on the pH of the drops. These may therefore be buffered to improve the stability. Care must be taken to ensure that the modifications to improve stability do not result in an irritant solution. An example of a buffered preparation is adrenaline eye drops; these contain a borate buffer to give a final pH of 7.4.

Light

Many drugs are adversely affected by light which may act as a catalyst to oxidative and hydrolytic processes. Drugs which are susceptible in this way should be stored in dark containers (amber glass) preferably in a light-proof cupboard. Phenylephrine and physostigmine come into this category.

Heat

Some drugs are unstable at high temperatures. This will prevent the use of autoclaving and possibly heating with a bactericide as methods of sterilization. In this case filtration or aseptic techniques may be necessary to ensure sterility. Cyclopentolate is an example of a drug unstable to heat. Proxymetacaine is unstable at room temperature if oxygen and light are present; it must therefore be stored in a cool place.

Hydrolysis

As eye drops are usually formulated in an aqueous medium, hydrolysis is a common problem. The only solution is to select constituents which are not susceptible to hydrolysis. This is not always possible, for example, with the local-anaesthetic group of drugs. Hydrolysis of a drug may be reduced by storing the drops in a cool place and in the dark.

Sterility

Eye drops should be sterile, and when they are prepared in multidose containers the vehicle employed should be bactericidal and fungicidal to minimize the risk of contamination during use.

Sterilization

The *PC* (*Pharmaceutical Codex*) lays down three methods of sterilization together with regulations for the treatment of apparatus and containers. Apparatus and containers should be thoroughly cleansed before use. The teats of eye droppers should be cleaned and impregnated with a solution containing the bactericide and other preservatives which are to be included in the final preparation. They should be stored seven days in this solution prior to use to allow equilibration to occur.

Autoclaving

The drug is dissolved in the vehicle containing the bactericide and any other ingredients. The solution is clarified by filtration and transferred to the final containers. These are sealed so as to exclude microorganisms and autoclaved at 115 °C, 10 lb/in^2 (69 kPa) pressure for 30 minutes.

Filtration

The drug is dissolved as for autoclaving but the solution is then sterilized by passage through special bacterial filters. After filtration the solution is transferred to sterilized containers using aseptic techniques. The containers are closed to exclude microorganisms.

Heating with a bactericide

The drug is dissolved in the vehicle containing the bactericide. The solution is clarified by filtration and transferred to the final containers. These are closed to exclude microorganisms and the drops sterilized by maintaining at 98–100 °C for 30 minutes.

Drops may be sterilized by any other method provided that the final product is identical in appearance, quality and composition with one prepared by the above methods. For any method used the final product must comply with the sterility tests laid down in the *PC*.

Prevention of contamination

Eye drops are sterile until the time they are opened, after which there is the danger of bacterial and fungal contamination. The formulation must be able to maintain the sterile state. Any bactericidal or fungicidal substance included in the formulation must be:

(1) Effective at room temperature or below.
(2) Non-irritant and non-toxic.
(3) Compatible with the drug, other substances and containers.
(4) Stable.

The *PC* recommends three substances for use as preservative agents: phenylmercuric nitrate or acetate, 0.002%; benzalkonium chloride, 0.01%; and chlorhexidine acetate, 0.01%.

These three substances are chosen because they are effective (within limits), non-irritant, non-toxic and stable. The final choice of preservative is governed by its compatibility with the active drug in the formulation and the purpose for which the drops are to be used.

Phenylmercuric nitrate, acetate

This possesses both antibacterial and antifungal activity. Although its activity against *Pseudomonas aeruginosa* is not very marked, it is the best of the three recommended preservatives and the agent of choice for fluorescein drops. The antibacterial activity is unaffected by changing pH but it may be reduced by the presence of anionic agents. The solution should be protected from light. There is the possibility that metallic mercury may deposit on standing. It should not be used in drops such as pilocarpine which are intended for long-term use as there is the danger of mercurialentis. Sensitization may occur to phenyl mercuric nitrate. As a preservative in eye drops a concentration up to 0.002% is used.

Benzalkonium chloride

This is a cationic surface-active agent. It is an effective bacteriostat and bactericide which is active against a wide range of Gram-negative and Gram-positive bacteria although the action against *Ps. aeruginosa* is rather slow.

By virtue of its surface-active properties, benzalkonium reduces surface tension and causes changes in membrane permeability.

These changes allow vital enzymes and other cell constituents to leak out of the cell. At the concentration used it is well tolerated by the eye, causing no irritation. Benzalkonium is incompatible with some anions including nitrate, salicylate, fluorescein and sulphonamides. The solution should be protected from light. As a preservative for eye drops it is usually used in the concentration 0.01%. Benzalkonium will also aid penetration of the cornea by drugs; this may be useful in the case of carbachol, or undesirable for local anaesthetics.

Chlorhexidine acetate

This is active against Gram-negative and Gram-positive bacteria with some activity against *Ps. aeruginosa.* It is unaffected by cationic compounds but is incompatible with anionic substances, for example, fluorescein. It should be protected from light and not allowed to come into contact with cork as this causes inactivation. Chlorhexidine acetate is usually used as a preservative in eye drops at a concentration of 0.005 to 0.01%.

Cetrimide

This is a cationic quaternary ammonium compound with an activity similar to benzalkonium chloride. It may be used for preserving eye drops at a concentration 0.005%. It may also be used for cleansing skin and in contact lens work.

Thiomersal

This has a similar action to phenylmercuric nitrate. It is bacteriostatic and fungistatic and is active against *Ps. aeruginosa.* For preserving eye drops it may be used at 0.01 to 0.02%. The precautions and incompatibilities are the same as for phenylmercuric nitrate.

Chlorbutol

This is usually bacteriostatic and active against Gram-negative and Gram-positive organisms. It will inhibit the growth of *Ps. aeruginosa* at 0.5% and is also active against fungi. It is stable to autoclaving at pH 6 or less but above this pH hydrolysis is increased by heat. It should be protected from light. A 0.5% solution is close to saturation point and crystals may deposit at low temperature.

Chlorocresol

This is a potent bactericide. It is used at 0.03 to 0.05% when it is predominantly bacteriostatic. Some people may find this concentration painful on instillation. The solution should be protected from light.

Tonicity

At one time isotonicity was considered desirable even if not of prime importance. Many eye drops are hypertonic and this may cause stinging on instillation. Hypertonicity can only be corrected by dilution which would result in dilution of the active drug, usually below its effective concentration. Hypotonic solutions may be adjusted to isotonicity by the addition of sodium chloride.

Containers

The *PC* lays down specifications for eye drop containers. Glass bottles should be amber, vertically ribbed (easily distinguished by touch from bottles with preparations for internal use), and made of neutral or soda glass provided that it has been treated to reduce the amount of alkali leached out by aqueous solutions (suitable plastic applicators may be used if they are capable of closure to exclude microorganisms). These containers should be fitted with a phenolic screw cap. The cap incorporates a dropper of neutral glass and a teat of natural or synthetic rubber. Teats should be able to withstand autoclaving, and should not release alkali or other harmful substances. Silicone rubber is preferred if benzalkonium chloride is used, as this reacts with the fillers used in the preparation of rubber. Alternatively, a complete dropper closure may be sterilized and supplied separately in a sealed package. In this case the bottle is closed with a plain phenolic plastic screw cap fitted with a suitable liner. The closure on bottles should be covered by a readily breakable seal.

Factors to be considered in the use of multidose containers

While eye drops in multidose containers are prepared with a preservative there is always the danger of contamination with resistant organisms or the accumulation of dead bacterial cells and their contents. Therefore, in an attempt to minimize this, the *PC* has made recommendations as to how long a multidose container should remain in use after opening. For home use a period of not longer than four weeks is recommended and eye drops supplied for home use should carry a label to this effect, unless there are other special storage recommendations. For hospital wards a separate container should be provided for each patient and for each eye if both eyes are being treated. They should be discarded not later than one week after first opening. In outpatient clinics and casualty departments opened containers should be discarded at the end of the day. Any patient who has undergone outpatient surgery should be treated with a separate supply of drops.

In operating theatres single-dose containers should preferably be used. If only multidose containers are available a previously unopened container should be used for each patient.

Eye drops of the *PC* are not intended for introduction into the anterior chamber during surgery. Drops for this purpose should be sterile but contain no preservatives.

An obvious way to avoid contamination is the use of sterile single-dose disposable containers. These contain a small quantity (0.3 ml) of sterile fluid (sterilized by autoclaving at 115 °C for 30 minutes) in a sterile plastic container. Not only are they sterile and disposable, thereby avoiding contamination, but they contain no preservatives. This lessens the likelihood of irritation and there is not the complication of incompatibilities in formulation. They may be preferable also for wearers of soft, acrylic contact lenses as these lenses are known to concentrate some preservatives, possibly leading to ocular irritation. Although these may be more expensive than traditional forms of presentation, they offer the advantages of sterility, stability and an increased shelf life. The range available is: atropine (1%), homatropine (2%), cyclopentolate (0.5 and 1%), hyoscine (0.2%), tropicamide (0.5 and 1%), phenylephrine (10%), pilocarpine (1, 2 and 4%), thymoxamine (0.5%), amethocaine (0.5 and 1%), oxybuprocaine (0.4%), lignocaine plus fluorescein (4 and 0.25% respectively), fluorescein (1 and 2%), rose bengal (1%), sulphacetamide (10%), sodium chloride (0.9%) and castor oil. Also produced but not available to the ophthalmic optician are: chloramphenicol (0.5%), gentamicin (0.3%), neomycin (0.5%) and prednisolone (0.5%).

Eye ointments

These contain the drug in a greasy base and are intended for application to the conjunctival sac or lid margin. The ointment is prepared from sterile ingredients using aseptic techniques. All apparatus to be used must be thoroughly cleansed and sterilized.

In all cases the drug is incorporated in the following eye ointment basis: yellow soft paraffin, 80%; liquid paraffin, 10%; and wool fat, 10%.

All the above ingredients are heated together and the molten base passed through a coarse filter paper into a clean container, the whole is then maintained at 160°C for one hour to ensure sterility. There are two methods of drug incorporation.

(1) As the base contains wool fat it absorbs water (up to 10%). Thus if the drug is water soluble, the sterile drug is dissolved in the minimum quantity of water and the solution sterilized by autoclaving or filtration and incorporated gradually in the melted sterile basis by aseptic technique. The finished ointment is transfered to the final sterile container which is closed to exclude microorganisms.
(2) If the drug is not water soluble but soluble in the ointment basis, the sterile medicament is finely powdered and thoroughly mixed with a small portion of the sterile melted basis. This mixture is incorporated with the rest of the sterile melted basis and transferred to the final container which is closed to exclude microorganisms. If the drug is not soluble in water or the basis, then it must be extremely finely powdered before incorporation to avoid irritation to the eye. After preparation, the ointment must comply with the sterility tests of the *PC*. Eye ointments contain no preservatives (bacterial or otherwise); however, it is difficult for bacteria to survive in non-aqueous conditions, therefore contamination is unlikely. There is little possibility of oxidation or hydrolysis. When the ointment is supplied for home use the patient should be told to avoid contamination. The ointment should be applied direct from the container.

In hospitals and clinics separate containers should be used for each patient.

Containers

These are small collapsible tubes of suitable metal or plastic or suitable single-dose containers. Containers should be as free as possible (consistent with good manufacturing practice) from dirt and particles of the materials used in their manufacture. Tubes, caps and wads should be sterilized before use, cork wads are undesirable as they may harbour fungal spores. The container should be closed with a screw cap covered with a readily breakable seal or the whole tube enclosed in a sealed plastic envelope.

Eye lotions

These are sterile aqueous solutions to be used undiluted in first aid. These are of two types, as follows.

(1) Sterile aqueous solutions containing no bactericide for first-aid or other purposes over a maximum period of 24 hours.
(2) Aqueous solutions containing a bactericide used for intermittent domiciliary use for up to 7 days.

The information on preparation and the standard for sterility applies to (1).

The *PC* recommends that the apparatus used in their preparation and final containers should be thoroughly cleansed before use. The medicament is dissolved in water, clarified by filtration and transferred to the container. This is closed to exclude microorganisms and sterilized by autoclaving. Alternatively, the solution may be sterilized by filtration, transferred to the final sterile containers and then closed to exclude microorganisms. Not more than 200 ml should be supplied in a container. Lotions should comply with the sterility tests of the *PC.* No preservatives are added, lotions are intended for short-term use only.

Lotions are also used for self-medication and are available in many proprietary forms. Some of these contain preservatives and may be used over a more prolonged period. Indiscriminate use at home should be discouraged as the use of a lotion may give temporary relief of ocular symptoms which require medical attention.

Containers

These should be coloured, fluted glass bottles. There should be an impermeable closure which must not contain cork, and be covered by a readily breakable seal. The use of non-fluted or colourless bottles is permitted if the fluted bottles are unavailable.

A number of eye lotions were included in the *BPC, 1963* but were subsequently deleted. The only one remaining is sodium chloride BPC 1973.

Sodium chloride

The eye lotion contains 0.9% sodium chloride in water and is isotonic with plasma. This lotion may be used in any situation where irrigation of the eye is required.

14

Legal aspects of sale and supply of drugs

A number of the preparations used by the ophthalmic optician contain substances which are regarded as potentially dangerous in one way or another. It is not desirable that these substances be freely available; their supply therefore is restricted to people who have a legitimate requirement for such substances. The requirement may be one of use in a profession, for example that of ophthalmic optician, or it may be to treat an illness when the supply is made to a member of the public by a pharmacist or on the prescription of a registered doctor, dentist or veterinary practitioner. Thus the legislation is designed to eliminate, or at least make difficult, criminal or injudicious use (for example of antibiotics leading to the development of resistant organisms) and to restrict access to those drugs with abuse potential, while not preventing the use of such drugs for their intended purposes.

The relevant Acts of Parliament are the Medicines Act 1968 and the Misuse of Drugs Act 1971. Non-medicinal poisons are dealt with separately by the Poisons Act 1972.

Misuse of Drugs Act 1971

This, together with the subsequent Statutory Instruments, deals with those drugs which are liable to abuse. The drugs covered, known as controlled drugs (CD), include opiates (such as morphine and heroin), the CNS stimulants (such as cocaine and amphetamine), the hallucinogenic drugs (such as LSD) and cannabis. The barbiturates, with the exception of those used for general anaesthesia, are to be brought within the scope of this Act from early 1985. The Act and Regulations cover the import, export, manufacture, sale and supply, and possession of controlled drugs, together with such matters as the safe custody of controlled drugs, the notification of and the supply to addicts. Under the terms of the Act, possession of controlled drugs is illegal unless obtained on a prescription or held by a member of a named profession in pursuance of that profession. The profession of ophthalmic optician is not named in the Act in connection with any controlled drugs and opticians may not therefore possess any such drugs in connection with their profession. It should be pointed out that the only controlled drug which might be of interest to the optician is cocaine for its properties as a local anaesthetic, mydriatic and vasoconstrictor. However, there are available alternative compounds which better meet the requirements of the optician with regard to these activities and there were no grounds for requiring the inclusion of opticians among those professions allowed to possess controlled drugs.

Medicines Act 1968

This comprises eight parts and numerous sections in total. Part III deals with the sale and supply of medicines and is of the greatest relevance to the optician. The remaining parts deal with matters such as the administration of the Act, manufacturing licences, pharmacies, containers and labelling, promotion of sales, the British Pharmacopeia and other publications, and miscellaneous and supplementary provisions which deal, among other items, with contact lenses and their solutions.

The general principle of the Medicines Act with regard to sale and supply of medicines is that this should only take place from registered pharmacies and should be by or under the supervision of a registered pharmacist. The application of this principle may be modified, and has been, in respect of some medicines where it is recognized that the hazard to health, the risk of misuse or the need for special precautions in handling is small, and where a wider sale would be of convenience and benefit to the purchaser. It is these medicines which constitute the General Sales List. The general principle applies to all other medicines and they are available to the public only from pharmacies with varying degrees of control exerted over their sale or supply.

The net result of the legislation is that medicines are in three categories: general sales, pharmacy only or prescription only. Those drugs categorized as general sales or prescription only appear in the relevant statutory order, namely the General Sales List Order and the Prescription Only Medicines Order. There is no statutory list of pharmacy medicines as such but any medicine which is not included in the General Sales List or the Prescription Only Medicine Order is automatically a pharmacy medicine.

General Sales List (GSL) medicines

Medicinal products on the General Sales List may be sold either from registered pharmacy premises or from other permanent retail premises, or in some cases automatic vending machines. The medicines must be sold in original packs and there may be restrictions on the pack size, strength of certain preparations, and so on. No eye drops or eye ointments are included under general sales regulations, even though the constituent may be totally free of restriction (for example, sodium chloride eye drops). There are no products of professional interest to the ophthalmic optician included in the General Sales List.

Pharmacy (P) medicines

Pharmacy medicines are those medicinal products which are not included in the GSL or the Prescription Only Medicine Order. Such medicines must be sold or supplied from a registered pharmacy and the sale must be by or under the supervision of a registered pharmacist. All eye drops or ointments which are not included in the Prescription Only Medicine Order are automatically pharmacy medicines, none may be included in the GSL. In this category are fluorescein drops (and papers), phenylephrine drops, Brolene drops and ointment, sodium chloride drops and naphazoline drops with a concentration below 0.015% (above this the product becomes prescription only medicine) and Optrex drops (the lotion is GSL).

Opticians are granted an exemption to the regulations regarding supply of pharmacy medicines and may in the course of their professional practice and in an emergency supply such preparations directly to their patients for use at home. It should be remembered that, under normal circumstances, the patient would be expected to obtain any pharmacy medicines needed from a registered pharmacy; the exemption should not be seen as a permit to opticians to undertake the regular supply of pharmacy medicines.

Prescription Only Medicines (POM)

These are listed in the Prescription Only Medicines Order and comprise those medicines where there is a known or potential toxicity, a danger of abuse and dependence (such as the controlled drugs) or a danger to the health of the community through injudicious use (as with antibiotics). Also included are all parenteral injections, including sterile water but excluding insulins for human use. Medicinal products containing POMs must be supplied from a registered pharmacy, by or under the supervision of a registered pharmacist and may normally only be supplied in response to a prescription from a registered doctor, dentist or veterinary practitioner.

Many of the preparations required by the optician contain substances which appear in the POM Order. Such preparations are available to the optician without difficulty because:

(1) The particular form or concentration of drug presentation may be exempt control even though the substance itself is POM.
(2) The ophthalmic optician is named in exemption orders in connection with certain named medicines such that the normal regulations regarding retail supply no longer obtain, and wholesale dealers may also supply the relevant medicines direct to the optician.

Drug form exempt

Products containing POM substances may be exempt from POM requirements on grounds of maximum strength in the preparation, maximum single dose, maximum daily dose or route of administration. For example, adrenaline and ephedrine as their various salts are POM but are exempt in preparations for external use (including eye drops). Naphazoline hydrochloride is POM but eye drops containing no more than 0.015% naphazoline hydrochloride are exempt. These preparations all become pharmacy medicines.

Medicines exempted in relation to ophthalmic opticians

The pharmacist is granted an exemption from the normal requirements with regard to sale and supply in respect of the drugs listed below, provided that the sale is subject to presentation of an order signed by a registered practising optician. The medicines in question are:

(1) Mafenide propionate.
(2) Sulphacetamide sodium, not more than 30%.

(3) Sulphafurazole diethanolamine equivalent to not more than 4% sulphafurazole.
(4) Atropine sulphate.
(5) Bethanecol chloride.
(6) Carbachol.
(7) Cyclopentolate hydrochloride.
(8) Homatropine hydrobromide.
(9) Hyoscine hydrobromide.
(10) Naphazoline hydrochloride, nitrate.
(11) Neostigmine methylsulphate.
(12) Physostigmine salicylate, sulphate.
(13) Pilocarpine hydrochloride, nitrate.
(14) Tropicamide.

The order may be presented by the optician and the supply made direct to him but it may also be presented by, and the supply made direct to, a patient or the patient's parents if the medicines required are to be used at home rather than in the optician's practice. The details required on such orders are:

Name, address and qualifications of the optician.
Date.
Name and address of patient (if applicable) and age (if under 12).
Name, form (say, ointment or drops) and strength of the medicine required.
Quantity required.
Purpose for which it is required (this may simply be 'for use in my practice' or it may be more precise if the supply is to a patient for home use).
Any labelling directions.
The signature of the ophthalmic optician.

The signed order should be retained by the pharmacist.

The ophthalmic optician is also granted an exemption with relation to the POMs listed earlier in that he may, in the course of his professional practice and in an emergency, supply the medicines in question directly to a patient for use outside the practice. There is no definition of 'an emergency', the decision rests with the optician and he must be prepared, and able, to defend his interpretation of 'an emergency' if necessary. As with the pharmacy medicines, it must be stressed that this extension of the optician's facility with drugs is not to be seen as a mechanism whereby he becomes a routine supplier of medicines, the usual procedure should be for the patient to obtain supplies from a pharmacist.

The pharmacist is also exempted the normal requirements for supply of POMs in respect of the drugs listed below when they are supplied to opticians for use in ophthalmic practice:

(1) Amethocaine hydrochloride.
(2) Framycetin sulphate.
(3) Lignocaine hydrochloride.
(4) Oxybuprocaine hydrochloride.
(5) Proxymetacaine hydrochloride.
(6) Thymoxamine hydrochloride.

and eye ointment containing oxyphenbutazone.

The ophthalmic optician may use this second group of drugs *only in the practice.* Under no circumstances may they be supplied to a patient directly, nor may a patient present a signed order to the pharmacist.

The optician may also obtain the drugs on either list for use in his practice by wholesale transactions. If the wholesale transactions involve a retail pharmacy then the signed order is still required; if a purely wholesale company is involved then a signed order may not be essential, although its use would seem preferable.

It should be noted that a number of items in the list, for example, sulphafurazole and bethanecol, are not presently available as formulations for use in the eye. Their inclusion in the list, however, ensures that should these medicines become available as eye drops or ointment no change in legislation is required to enable the optician to obtain them. For example, thymoxamine was listed as one of those POMs available to the optician a number of years before there was a suitable formulation. Once this was produced there was no delay in opticians being able to obtain thymoxamine because of its prior inclusion in the list of POMs to which exemptions apply.

Appendix

Passage of drugs across the cornea

The majority of drugs used by the optician are weak bases employed as their water-soluble acid salts. *Figure 2.1* shows phenylephrine hydrochloride as a specific example. The general equation for the equilibrium between ionized and un-ionized forms of a weak base may be written as shown in *Figure A.1.*

$$\underset{\text{Ionized}}{RNH_3^+} + Cl^- \rightleftharpoons \underset{\text{Unionized}}{RNH_2} + H^+ + Cl^-$$

Figure A.1. Dissociation of a weak base in solution

The ratio of ionized and un-ionized forms present in solution depends on two factors: the pKa of the drug and the pH of the solution in which it exists. This ratio may be determined from the Henderson–Hasselbach equation (*Figure A.2*). The pKa is an unchanging property of the drug and is the pH at which it is 50% ionized (that is, the concentrations of the ionized and un-ionized forms are equal).

$$pH = pKa + \log_{10}\frac{[base]}{[acid]}$$

in this instance

$$pH = pKa + \log_{10}\frac{[RNH_2]}{[RNH_3^+]}$$

Figure A.2. The Henderson–Hasselbalch equation

Considering the example of a weak base pKa 9.4 (approximately that of atropine) prepared in a formulation of pH 6.4 as its acid salt. Then from the equation (*Figure A.3*), the ratio of the

$$6.4 = 9.4 + \log_{10} \frac{[RNH_2]}{[RNH_3^+]}$$

$$\log_{10} \frac{[RNH_2]}{[RNH_3^+]} = \bar{3}$$

$$\frac{[RNH_2]}{[RNH_3^+]} = \frac{1}{1000}$$

Figure A.3. Calculation of degree of ionization of a weak base (pKa 9.4) in a solution of pH 6.4

un-ionized to ionized will be 1:1000. If the pH surrounding the drug is 7.4, as in tears, the ratio will be 1:100 (*Figure A.4*). At the surface of the cornea, the pH will be somewhere between 7.4 and 6.4 depending on the buffering ability of the tears. The proportion of the un-ionized form in solution will thus be somewhere between 1:100 and 1:1000. It is this un-ionized lipid-soluble form which crosses the epithelium.

$$7.4 = 9.4 + \log_{10} \frac{[RNH_2]}{[RNH_3^+]}$$

$$\log_{10} \frac{[RNH_2]}{[RNH_3^+]} = \bar{2}$$

$$\frac{[RNH_2]}{[RNH_3^+]} = \frac{1}{100}$$

Figure A.4. Calculation of degree of ionization of a weak base (pKa 9.4) in a solution of pH 7.4

Consider the equation (*Figure A.4*) where the base is one part in 100 parts un-ionized. When this one un-ionized part crosses the epithelium into the stroma, it will, itself, re-ionize according to the equation (*Figure A.5*). Assuming the pH in the stroma to be 7.4, the ratio of ionized to un-ionized will be 100:1. The ionized form, being water soluble, will cross the stroma to the endothelium. Here a new equilibrium will be established between the two forms and the 1 part in 100 which is un-ionized will cross the endothelium into the aqueous.

The whole situation is a dynamic one and once any equilibrium is disturbed, as for example by the un-ionized form crossing the epithelium, the equilibrium will be re-established, in this case by dissociation of RNH_3^+. This will thus provide more un-ionized form to cross the epithelium, and again disturb the equilibrium. This re-establishment and disturbance of the equilibrium will take place at each of the three interfaces. In this way the drug will cross the cornea following a concentration gradient.

$$RNH_3^+ \rightleftharpoons RNH_2 + H^+$$

$$\log_{10} \frac{[RNH_2]}{[RNH_3^+]} = \bar{2}$$

$$\frac{[RNH_2]}{[RNH_3^+]} = \frac{1}{100}$$

Figure A.5. Calculation of degree of ionization of a weak base (pKa 9.4) in the stroma at pH 7.4

If a drug is completely ionized it will be unable to penetrate the epithelium. If the pKa of a weakly basic drug is low the ratio of un-ionized to ionized drug is high and there is therefore a higher proportion of this form available to cross the epithelium initially. A small change in pKa causes a large change in these ratios.

Figure A.6 shows the effect of changes in pKa in a series of local anaesthetic drugs. Thus, as the pKa rises from 7.9 for lignocaine through 8.5 for amethocaine to 8.9 for procaine, the ratio of

	pKa	$\frac{[RNH_2]}{[RNH_3^+]}$ at pH 7.4
Lignocaine	7.9	1/3
Amethocaine	8.5	1/13
Procaine	8.9	1/32

Figure A.6. The effect of changing pKa on the degree of ionization of a series of local anaesthetics at pH 7.4

un-ionized to ionized form in a solution at pH 7.4 changes from 1:3 to 1:13 to 1:32. Because of its higher pKa a larger proportion of the procaine molecule is ionized; conversely, because of its lower pKa a smaller proportion of the amethocaine molecule is ionized. This is one reason why procaine is less efficiently absorbed and therefore less effective as a corneal local anaesthetic.

The pKa of a drug cannot be changed but the pH of the eye drop formulation may be changed, within limits, to encourage a higher proportion of un-ionized drug at the cornea. Thus, as the pH of the formulation is raised, the degree of ionization of a weak base will be reduced. The pH cannot be increased over the full range as too drastic a change from 7.4 (the usual pH of body fluids) will result in irritation of ocular tissues and possible ocular damage. Changes in pH may also affect the solubility and stability of the drug. The pH must therefore strike a balance to give stability, reduced ionization and minimal irritation.

Further reading

BOWMAN, W. C. and RAND, M. J. (1980). *Textbook of Pharmacology*, 2nd edition. Oxford: Blackwell

BRITISH MEDICAL ASSOCIATION AND THE PHARMACEUTICAL SOCIETY OF GREAT BRITAIN. *British National Formulary*. (Twice yearly publication)

DAVIES, D. M. (1981). *Textbook of Adverse Drug reactions*, 2nd edition. Oxford: Oxford University Press

DAVSON, HUGH. (1980). *Physiology of the Eye*, 4th edition. Edinburgh: Churchill Livingstone

ELLIS, PHILIP P. (1981). *Ocular Therapeutics and Pharmacology*, 6th edition. St Louis: C. V. Mosby

FOSTER, R. W. and COX, B. (1980). *Basic Pharmacology*. London: Butterworths

FRAUNFELDER, F. T. (1982). *Drug Induced Ocular Side Effects and Drug Interactions*, 2nd edition. Philadelphia: Lea and Febiger

GOODMAN, L. S. and GILMAN, A. (1980). *The Pharmacological Basic of Therapeutics*, 6th edition. London: MacMillan

O'CONNOR DAVIES, P. H. (1981). *The Actions and Uses of Ophthalmic Drugs*, 2nd edition. London: Butterworths

PARR, JOHN. (1982). *Introduction to Ophthalmology*, 2nd edition. Oxford: Oxford University Press

PEARCE, MAUREEN. (1982). *Medicines and Poisons Guide*, 3rd edition. London: The Pharmaceutical Press

SASCHSENWEGER, RUDOLF (1980). *Illustrated Handbook of Ophthalmology*. Chichester: John Wright and Sons

STONE, J. and PHILLIPS, A. J. (Editors) (1980). *Contact Lenses*, 2nd edition. London: Butterworths

Index